Willis P. Hazard

How to Select Cows

Or, the Guenon system simplified, explained and practically applied

Willis P. Hazard

How to Select Cows
Or, the Guenon system simplified, explained and practically applied

ISBN/EAN: 9783337182946

Printed in Europe, USA, Canada, Australia, Japan

Cover: Foto ©Lupo / pixelio.de

More available books at **www.hansebooks.com**

HOW TO SELECT COWS;

OR

THE GUENON SYSTEM

SIMPLIFIED, EXPLAINED, AND PRACTICALLY APPLIED.

BY

WILLIS P. HAZARD,

Secretary of the Pennsylvania Guenon Commission; President of the Chad's Ford Farmers' Club; a Vice-President of the American Dairyman's Association; Lecturer upon Agriculture at the Delaware State College, &c., &c.; Author of Treatises "On the Jersey, Guernsey, and Alderney Cow," and "On Butter and Butter-making;" "The Annals of Philadelphia," &c.

WITH NEARLY 100 ILLUSTRATIONS
Photographed from Guenon's engravings.

PHILADELPHIA:
J. M. STODDART & CO., 1018 Chestnut Street.
1882

PREFACE.

The want has long been felt for a hand book which would simplify and explain the invaluable discovery of Guenon, to enable any one to select good stock. There can be no doubt if this discovery is made to be easily acquired, that millions of dollars would be saved to the community by the improvement of herds and a consequent reduction in the price of bovine products, on account of the increased yield and the lower cost of raising it.

The State of Pennsylvania, in 1878, appointed a commission to test the system and report upon it. As a member of that commission, we found there was with many a superficial knowledge of the subject, with others enough acquaintance with the system to destroy their faith in it, and with nearly all a desire to obtain sufficient practical knowledge of the system to enable them to judge understandingly and to practice it.

With a view to fill these wants, we have undertaken the explanation of the system in the following pages to enable all to fill up their measure of knowledge of the system, so that the superficial shall become thorough, the doubting acquire new faith, and all see its merits *the more they practice and apply it.*

We have accordingly given a sketch of M. Guenon and the progress of his discovery; some extracts from his preface explaining his views; an explanation of his system of escutcheon marks; a description of the various escutcheons and their indications of value and quantity, and directions how to practically apply them, together with the report of the Pennsylvania Guenon Commission.

Believing that we have thus presented a comprehensive view of this discovery, we trust every one into whose hands this work may come, will patiently, book in hand, go into the farm-yard and judge of the value of his stock by the rules here set forth, compare the results with his individual knowledge of his stock, and fairly estimate the value of the system.

The illustrations of the escutcheons are photographed from Guenon's drawings in his last revised edition.

WILLIS P. HAZARD.

MAPLE KNOLL, WEST CHESTER, PA., *September, 1879.*

LECTURES ON THE GUENON SYSTEM.

The author of this work having been invited to lecture a number of times before agricultural societies, and being constantly in receipt of letters of inquiry about repeating his lectures, takes this public opportunity to announce that he will make arrangements to repeat his lecture on the Guenon system, illustrated with a series of large drawings. Correspondence is solicited with officers of agricultural societies, granges, agricultural fairs, colleges, farmers' clubs, and dairymen's associations. His principal object being to disseminate widely a knowledge of a system of such great value to his brother farmers, the terms will be satisfactory.

At each lecture practical illustrations and instruction is given in the barn-yard or field. Address him at

WEST CHESTER,
CHESTER COUNTY,
PENNSYLVANIA.

LECTURE UPON THE CHANNEL ISLANDS, THEIR PEOPLE, AND THE CATTLE.

The author having recently spent several months in Guernsey and Jersey for the purpose of studying the habits of the people, viewing the scenery, and acquiring a knowledge of the agriculture, and the breeding of their cattle, has embodied the results of his visits in a lecture, which he is now prepared to deliver before agricultural and other associations.

LECTURES UPON AGRICULTURE.

The author having been appointed Lecturer upon Agriculture to the Delaware State College, at Newark, Del., will repeat all or part of the course to farmers' institutes, colleges, schools, &c. The lectures are popular in character, and not too scientific for general comprehension.

GUENON'S INTRODUCTION
TO HIS NEW REVISED EDITION.

Error is propagated with the rapidity of lightning; before it every obstacle disappears, and popular favor seems to welcome it. Truth, on the contrary, is received with indifference, often even with doubt, suspicion, and distrust. Indeed, how often have we not seen the author of a discovery which, having been accepted and realized ought to have advanced the public good and increased the general welfare, come into contact with the hatred, the ignorance, and the envy, and thus become the laughter of fools and the jest of the wise. To some the inventor seemed without good sense; to others an ignoramus. Too feeble to struggle against all, he died broken hearted, and left to his powerful antagonists the glory of having, perhaps for ages, buried his discovery, and to those who bring to perfection cities and fields the deprivation of a good up to that time unknown.

If more happy than those martyrs for a new idea, I should reach, at last, that which concerns me, after twelve years of incessant struggles, to cause the truth to appear to the eyes of all, I should have nothing more to desire. Nothing would remain for me, except to bless the generous hearts which shall have aided me in triumphing over routine and error; then on quitting this world, only to bequeath the worthy men who have so bravely encouraged and seconded my efforts, the task of simplifying my discovery, and rendering my method popular to cause the analytical knowledge of cattle to penetrate even into the most obscure hamlet, and while dividing thus with me the glory of having done this good, their names will be held in grateful remembrance by future generations; such has been the idea which has guided every moment of my life, all the efforts of my mind.

For nearly twelve years, since I have given my method to the public, through a first edition of my "Treatise on Milch Cows," the savants and the practitioners have been greatly prepossessed with it. When they have seen me make a successful application before them of my system, by a single inspection of animals which I saw for the first time, they have expressed a lively surprise.

In the vegetable kingdom, skillful nurserymen have distinguished more than eighty different orders of pears of summer, of autumn, and of winter; each of these orders has its distinctive characteristics, as many for the shape and the taste of the fruit, as for the time of ripening. And when a tree-grower or an amateur is sufficiently skillful, he distinguishes marvelously all these species one from another by a single inspection, and at whatever time of year it may be. He knows equally well what exposure it is necessary to give to each of them to obtain exquisite fruits.

My first studies had been directed toward arboriculture. I have practiced with my father during many years. My principal occupation was the cutting of trees, grafts, both splits and bandages, and by studying vegetables, I had acquired the idea of and an insight into classifications.

I was better prepared thus for my work of classifying the bovine race, a work which no one had ever tried, either theoretically or practically.

My classification of the characteristic signs, embraces all the races of France and other countries, without distinction of sex or age.

Unknown, up to this day, although they have always existed, these signs have escaped all the world, even the sagacity of the most celebrated painters, as well as that of veterinary doctors of the highest reputations of all times.

The appearance of my method should mark an era, for it opposes and overturns all the prejudiced routines according to which people have practiced up to this time.

It opens a new era in an art in its infancy, in a science whose first principles even were unknown. I should then expound it with the greatest detail.

This method is of the greatest simplicity, whatever has been possible to be said of it, and whoever will become thoroughly familiar with the escutcheon of the first order of each class, will be able to judge of all.

Escutcheons are ten in number. They extend, according to their class, from the centre of the four teats to the level of the upper extremity of the vulva, and may extend in breadth from the middle of the hinder surface of one leg, to the middle of the hinder surface of the other. By their form or configuration, escutcheons characterize and distinguish the ten families which together constitute my classification. Behold, then, to what is reduced, in reality, this pretended immense complication.

A special figure, placed at the end of each class, serves to indicate mongrel animals.

Each of the *classes* or families is characterized by an escutcheon of fixed form, always similar to itself, while one does not get out of that class or that family, but variable in the dimensions of its surface. That dimension or that surface should be estimated by square centimeters, but that would be too complicated for the practical man; since it depends on the size of the individual, it is estimated by the limits of the escutcheon placed on the hinder part of the animal. The extreme limits are the hams, the interior surface of the legs and vulva. The surface of the escutcheon, of which the extent varies, has permitted me to divide each class or family into *six orders*, for each one of which I assign, in taking account of the shape, the quantity, the continuation, and the quality of the milk.

The escutcheon of the first order is the most developed; is also the best marked. The escutcheon of each of the five other orders is similar in form to that of the first order. It is, in some sort, only a proportionate reduction—a diminutive. It is the escutcheon of the first order, with the dimensions reduced or brought within less extended limits, reaching no longer the hock nor longer covering the interior of the thighs, nor yet reaching up to the vulva, remaining consequently at a distance greater or or less from these boundaries.

I have added to this new edition—

1st. Two new classes, sub-divided also into six orders, (the Left Flanders and the Double Selvage.)

2d. Two varieties of escutcheons, having some similarity with the others.

3d. Finally, the classification of the bull.

These three additions, unpublished until now, complete and generalize the system of characteristic signs, by which one can prove the absolute and relative superiority or inferiority of each individual of the race.

These new forms of escutcheons were known to me at the time of the publication of my first issue, and which I had already announced; but they occurred so rarely in the races which were familiar to me, that I thought they were not worth publishing.

But, now, since I have traveled so much, not only in France, but in foreign countries, I have convinced myself that these classes occur much more commonly in certain races than I had thought at first. I have felt the necessity of putting them in my method, and have given them their proper place.

In respect to the two new varieties of escutcheons, they are like an appendix to the classification, and characterize the product of crossing between different classes.

To state precisely their signification and to value their corresponding milk product, it is necessary to compare these escutcheons with the order of the class to which they are the most analogous.

When I shall have described the different families of true cows, as well as their division into orders, the yield or the quantity of milk, their butyraceous qualities, and the greater or less period of its duration of yield during gestation, I will pass to the bastard cows, which, though perfectly similar in form and color to others, differ essentially from them, for they lose their milk as soon as they are pregnant.

This close resemblance is a source of errors to the most practiced judges.

Thus have I wished in the description of classification, to point out precisely the distinctive signs by the aid of which one can easily recognize them. After the study of bastard cows, comes the chapter of bull re-productors. I have made plain, that in the classifications of bulls, I have reduced to three the numbers of orders of each class, in order to bring the application of the method to the most simple expression. The first will comprehend all the bulls, the good re-producers; the second, the re-producers of middling quality; the third, the bad re-producers. I mean by bad, those in which fails the ability for the transmission of the lactiferous qualities. As one sees, the characteristic signs with the males, as with the females, have a significant value of the highest importance. With the bull, they portray the re-productive qualities, and with the cows the lactiferous qualities. The observers who will apply my system of one kind, as rigorously for the males as for the females, will observe in the passage of one order to the other, the same scale of proportion that this established in the classification of the cows. Although the classification bears more on the lactiferous or re-productive properties than on the others, it is important to take in consideration all the other qualities that the individuals can and ought to possess to be of an irreproachable organization.

The cows of the first and second order of each class, in all the races, will always give in the same country, a greater abundance of milk than those of inferior orders. To recognize the lactiferous produce of cows, whatever may be their class or the locality that they inhabit, it suffices simply to know the quality of the food which makes the habitual nourishment of the cows in the place where they are kept.

In following in his appreciation, the degree of superiority or of inferiority of the escutcheon, one will judge close upon the daily quantity of milk that all the cows of the same country are apt to give, for one will know then in what proportion all the figures of the classification should be modified. A milk cow ought to be neither too fat nor too lean, to give her maximum of milk. All confinements in a period of thinness is prejudicial to the habitual produce. Even when the animal would have recovered her strength, she will not recuperate so as to restore the quantity of her milk; that can take place only after a year, and by means of a new calf. A great milk cow, whatever may be her aptness for fattening, and her condition of fat at the time of calving, becomes thin about fifteen or twenty days after calving; the time of her rut is therefore less near than that of a poor milk cow, because her vital forces are weaker. Witness the quantity of her yield, which is only that of a cow of medium product.

One can compare a milch cow to a fruit tree, which gives more fruit this year than the next. When the sap of the tree carries vigor to the development of the fruit, the growth of the wood remains nearly stationary. When, on the contrary, the tree gives but little fruit, the sap turns to the profit of the wood, to give, after a repose of several years, a greater quantity of fruit, and to continue thus by alternative successions.

It is the same with the cow, for it is seldom that her produce keeps the same during three consecutive years, for the reason that, when the nourishment absorbed by her turns to the profit of the milk, the milk is more abundant; when, on the contrary, the nourishment goes to fat, the milk diminishes.

The variations in the milk quantity should be justly attributed to the influence of atmospheric circumstances of the seasons, which react on the quality of hay and fodder in augmenting or diminishing the nutritive juices of the food.

Cows which are fed in good pastures surpass the product which I have assigned to their class and their order, while those which are in poor and wet pastures have necessarily inferior produce, unless the latter have in the stable nourishing food, more abundant and more succulent than they are able to get for themselves out of doors.

If, for example, the well-fed cows, or those grazing on rich pasture lands, should give as much as twenty to twenty-five quarts of milk per day; these same cows, taken and fed on poor pasture, will give only about ten or twelve quarts.

If, on the contrary, one takes the cows raised on a poor soil, transfers them to rich pastures, the milk produce of these same cows will be superior to that they gave in their original lands.

My readers should well understand that in the valuations of my classifications that I have not pretended to assign a rigorous and absolute amount. I have been only able to give an approximate figure to each class and to each order, adopting the medium limit of the ordinary amount of the different breeds of various localities.

The atmosphere, the care, and the different foods of each country, all these different things exercise upon the animal, an influence favorable or unfavorable, according to the nature of the soil.

There are many other circumstances which should be considered, and which would disturb the harmony of the figures of my valuation and the normal quantity. Such are, for example, the case of sickness, accidents, &c. That is the reason I have adopted, in determining the quantity of cows of each order, a medium figure, such as is shown in the classification.

I will also observe, relative to those animals to which I assign approximate weight in the course of this work, that, following the customs of commerce, of sale, and of butchers, this weight is dead weight, the animal being deprived of the skin, intestines, head, feet, &c.

If, contrary to custom, I had acted otherwise, and had made the calculation for the animal on the hoof, the figures given by me would present a great difference, which would increase according to the amount of fat, sometimes to double the weight.

The discovery which I have made of the value of the escutcheon is designated by the contrary direction of the hair, and which had escaped the attention of every one, even those most interested in gaining the knowledge of it. It is necessary also to avow the effect produced by the change of direction of the hair is not glaring on the animal. It is merely a difference of luster, and the gloss on the surface of the escutcheon and the part of the skin surrounding it. The hair of the escutcheon is finer, shorter, more furry, and more silky. Its appearance, at the first glance, makes one think this part of the animal has been shaved. Compared with the ordinary hair, the skin of the udder appears to be more designed to be quicker seen on the part where appears the escutcheon.

All animals of the bovine species, without excepting even wild animals, are marked with an escutcheon, large, small, or medium, regular or irregular. Their characteristic sign is transmitted with the generating germ.

I have not thought it necessary to say much on that portion of the escutcheon which extends on the stomach of the beast towards the navel. This addition has been thought useless. Enough is shown of the escutcheon when she is standing.

In order to see well the escutcheons with all the fullness which my sketches give them, it must be supposed that the udder of each cow is seen at its greatest plenitude of milk, such as would separate the hind legs to the greatest extent. In this way the escutcheon is seen as if the entire skin of the animal was placed flat, or as if the envelope of the milk bearing apparatus formed a plain surface, on which are drawn the elevations, the depressions, and all that is not visible to the eye, without the aid of hands or of movement of the cow, both that which is hidden at the further side and in the folds of the udder and of the thighs of the animal on foot.

In order to examine and to distinguish perfectly the escutcheon, one should place himself behind the animal and make it advance some steps, in such manner that the movements which it makes in walking should show, one after another, the parts which one needs to see.

One can also, in passing the nails over the space occupied by the escutcheon and leading the hand downward from above, in a manner contrary to the rising hair, and ruffling it, recognize without difficulty its form and its extent.

Theoretical explanations are always abstract and diffuse in their development. My method may at first appear difficult and complicated, which, indeed, pretended savans have chosen to affirm. Nevertheless it is not so, and in order to comprehend it, it is sufficient to study it. It is with this as with everything else, to know it is necessary to study and to practice.

The beautiful art which I am about to explain to agriculturists is most easily acquired. Its technical dictionary is composed only of certain words, of which the readers should, first of all, know perfectly the precise signification.

These words are *Escutcheons, Epis or Tufts ascending*, and *Epis or Tufts descending*. After he knows perfectly the different forms and the importance of these characteristic signs, he will know the whole subject as well as I do myself.

The Epis or Tuft, as one will see, participates with the escutcheon in the distinction of the orders—it multiplies the sub-divisions. It seems at the same time to complicate my method and to render it less accessible; but I have not felt myself at liberty to omit it, since it has an incontestible and important value.

If, among certain animals, the form and extent of characteristic signs are not exactly those of the drawings, but a sort of intermediate between the characteristic signs of two classes, he who applies the method should approximate them to the drawing of the classification from which they differ the least, and from that deduce the probable value.

To render my work perfectly clear, I had to enter into the developments very much in detail. Nevertheless, so extensive are these details that I believe I have given neither too many nor too few, and have confined myself simply within the limits of the possible, the indispensible and the useful.

And now, whoever my opponents may be, I proclaim boldly and without fear, that the escutcheon is the only incontestible characteristic sign that can enable one to discern, by simple inspection, the aptitude for milk production of each animal.

All animals of the bovine species in good state of health, to which no accident has happened, and whose escutcheons are of the first orders of each class, will manifest always, and without exception, as much for the production of milk as for generative ability.

Beauty of form, to my thinking, represents but an ideal, and although one ought to take it into consideration, it is a simple accessory without value of its own, when the question is that of the production of milk.

May I have been able to justify by this work the fruit of the experience of my whole life, the honor done me by many agricultural societies in admitting me to their membership, and by the government which has shared the expense of this new edition, with the two-fold purpose of encouraging my efforts and facilitating the propagation of my method.

GUENON'S METHOD OF JUDGING OF THE VALUE OF STOCK.

Fifty years ago there was dawning upon the world the first ray of a great discovery. A star was rising in the agricultural world, which was about to shed new light, and like many other valuable discoveries, it was made by one among the lowly, and partly by chance. The author of this new discovery has said, "Error flies with the rapidity of lightning, all obstacles vanish before it. Truth, on the contrary, is admitted coldly, often even with doubt, suspicion, and distrust." It is owing partly to this, partly to the fact that this new light was given to the world when the mind of farmers were not ready to *receive* new ideas of progress as they now *seek* them, and much to the fact that it was the invention of a foreigner described in a foreign tongue. True a translation of it was made through the medium of an American monthly magazine of agriculture; but it was one of limited circulation. At that time the number of periodicals devoted to that interest was few, and such new and important questions were not throughly discussed and the knowledge of them placed in every farm-house in the land, as it is at the present day. Shortly after the appearance of M. Guenon's treatise in the magazine, it was reprinted in book form, and received the large circulation of sixty-five thousand copies, between that time and now, and the book most probably sells better to-day than it did then. By many who procured that book the subject was studied, and advantage taken of its revelations, being stored away in the reader's mind for actual practice. By the great majority it was read, but not studied; driven from it by the apparent complications of the system and the two hundred sub-divisions of it; by many, perhaps, it was attempted to be put into practice, but without their having given the subject that close investigation which was needed to prove the system correct. It was mostly by this class of persons, because the system was not found to be infallible, that it was denounced and given up, even by men otherwise intelligent; as if anything human could be infallible. Thus it is that by the ignorant its revelations were received with incredulity, and by many of the intelligent with doubt; but to the earnest seekers after practical information, it has unfolded a mine of wealth, and they have proved the system by continuous experience, and found it to be the most reliable mode of judging of the value of every member of the bovine species.

It was a happy thought that suggested itself to the Pennsylvania State Board of Agriculture, to have the system tested by uninterested parties. But extremely difficult, it was, to obtain persons to make the test. For those to whom application was made declined it on various grounds, principally because, as Guenon himself has stated in his latest edition, many pretended savans would endeavor to throw ridicule upon it; many others would identify the gentlemen making the tests with it, as if it was their system that they were testing; while not a few still more narrow-minded, would think they were trying to humbug them. Thus it was difficult to fill the places, which offered neither honor nor profit.

It will be seen, by these extracts, that the Governor appointed three experts to test the system. This they did in the summer of 1878, examining two hundred cows, jotting down their opinion of the yield, quality, and time of each of them, and afterwards printing them alongside of the reports of their owners, so that the public could form their own estimate of the results of the examinations of the commission. They are here re-printed, to show how it was carried out. Particular attention is called to the examinations *of the blanketed cows* in Thomas Gawthrop's herd.

On M. Guenon and his System.

It is proper we should inquire into M. Guenon, and the origin and development of his system.

Monsieur François Guenon, a husbandman of Libourne, in France, was the son of a gardener, and followed for sometime his ancestor's trade. He seems to have had a mind above those in his position. As we look at his portrait, he appears to have a clear eye, a cool head, great determination, firmness of character, a well-balanced mind, and with it all, a vigor of constitution which buoys him up, and enables him to over-ride obstacles. He says himself, he was of an observant turn of mind, fond of comparing things, and deducing consequences from what he learned by observation and comparison, particularly from the Book of Nature. Young, ardent and healthy, with the vivacity of his race, he felt himself destined for better things than those a gardener's life would insure him. What wonder then that his eye was keen to see, his mind to grasp and analyze any new turn of thought that chance might throw in his way.

Like most self-made men, who have made their mark in life's pilgrimage, he set himself to work to improve himself—to acquire that which would expand his mind, and fit it to receive any new inspiration, and be able to develop it. He studied the works of the best writers on botany and agriculture; and applied his knowledge by following up all the ramifications of the vegetable kingdom, and studied their external signs, that distinguish the different sorts, and ascertained their qualities and productiveness.

In France, they have few fences, and the cattle of a neighborhood are driven to the grazing ground, and herded together, and, in turn, members of each or several families, (the younger portion,) are put to watch that the cattle do not stray out of bounds. Such companionship with their stock makes the owners fond of them, and they are treated as pets, and become very docile. When young Guenon was about fourteen years of age, he would drive their cow to graze. His cow he was very fond of, and could identify her among any number. She was a good milker.

The Escutcheon or Mirror.

In his authorized account of the discovery and perfection of his system, Guenon uses the following language: "When fourteen years of age, I used, according to country custom, to drive our only cow to the grazing ground. I was very fond of her, and could have identified her among ever so many. One day as I was whiling away the time in cleaning and scratching my old companion, I noticed that a sort of bran or dandruff detached itself in considerable quantities from certain spots on her hind parts, formed by the meeting of the hair as it grew in opposite directions, which spots I have since called *ears*, from the resemblance they often bear to the bearded ears or heads of wheat or rye. This first attracted my attention, and I recollected having heard my grandfather say that it was probable that there were external marks on cows whereby their good qualities or their defects might be known—just as we judge of the vital force of a plant and its qualities by means of its leaves and lines in its skin. Reflecting on the subject, I arrived at the conclusion that if in the vegetable kingdom there exists external signs, whereby the good and the bad qualities of a plant can be positively known, there ought to exist in the animal, or its kingdom, also, marks whereby we may judge, by inspecting an animal, of its qualities, good and bad, and I thought I had discovered one of these signs. I sought the bearded ears or quirls, and scratched those spots in quest of dandruff, the abundance or scarcity of this being what first engaged my attention. Every new cow was compared with my own as a standard, and her superiority, equality, or inferiority determined in my own mind. In the course of the comparisons thus instituted by me, with reference to the dandruff alone, which was at first the only thing that governed me, I had occasion to remark that great diversities existed among cows in respect to the shape of the bearded ears (quirls) which produced the dandruff. This suggested a new train of reflection and observation, which resulted in my becoming convinced that these *shapes* were the signs by which to distinguish cows, and to know the good and bad qualities of every individual among them."

In his original plan, Guenon divided these different shapes into eight classes, each of which was sub-divided into eight orders. As he progressed in his investigations, he afterwards added two more classes, and reduced the orders to six in each class. These he supposed would cover all cases which might come up for examination. He also divided cows into three grades, which, in accordance with their *size*, he styled high, low, and medium. From this it will be noted that Guenon, in classifying cows, was governed first by the class, second by the order in the class, and finally by their size. These *classes* he divided and named as follows:

1st class, or Flanders.	6th class, or Double Selvage.
2d " " Left Flanders.	7th " " Demijohn.
3d " " Selvage.	8th " " Square Escutcheon.
4th " " Curveline.	9th " " Limousine.
5th " " Bicorn.	10th " " Horizontal.

The ten orders in each of these classes were simply designated by their appropriate numerals. Each *class* was better than the succeeding one, and each *order* better than the following one of the *same* class, but might be better than the preceding order of the *next* class.

Of this seeming multiplicity of classes, orders, and sizes, Chalkly Harvey, one of the commission appointed to test the system, writes thus:

"Now this may seem somewhat discouraging to your readers, but with all due respect to Guenon, to whom all honor and praise should be accorded for his brilliant

Imported Jersey Cow BLACK BESS.

Imported Jersey Cow TIBERIA.
Belonging to C. L. Sharpless, Philadelphia.

SHAPES AND SIZES OF ESCUTCHEONS.

discovery, I think that it may be so simplified that every farmer, dairyman, and dealer can learn it all in a short time, and may find the study quite interesting. I began it laboriously, supposing that a mastery of all the details was necessary to make it of any use, but more than twenty years of constant application in practice has simplified it to my mind, and has added a little, I think, to the original discovery. The substance of Guenon's discovery is that the milking qualities of any cow, of any breed, are indicated by an outward sign that all may see and easily understand. The hair on a cow, as on other animals, grows downward on the hind-quarters, but there is an exception to this rule on the back part of the udder, where it usually grows upward. The first lesson for a beginner is to notice this fact. Let him stand behind a quiet cow, and rub the hair on the udder both ways until he sees or feels just what I mean. Guenon called the surface that is covered by this upward growth the escutcheon; others have called it the milk-mirror; but this is no improvement in any respect, and I shall name it as Guenon did, for there is no real objection to that name, and there is serious objection to making confusion by calling the same thing by different names. The escutcheon, then, is that surface on the cow's udder where the hair grows upward. But it is not confined to the udder, it extends upward above the udder, often to the vulva, and outward upon the thighs on both sides of the udder. (See Flanders cow, class first, order first.) These escutcheons are different in size, in shape, and in quality, (quality means the quality of the skin, and of the hair growing on it,) and these differences indicate the different milking qualities of the cows, including quantity and quality of milk, and the length of time they will give milk after being with calf. On the edges of the escutcheon where the upward and the downward growths of hair meet, a feather is formed, and this is most conspicuous on the back part of the thighs where escutcheons extend that wide. If the hair is long, as it generally is in winter time, the observer can define the limits of the escutcheon bottom by applying his hand, and smoothing the hair to its natural place. He will now perceive that the hair on the escutcheon is shorter and softer than elsewhere, as well as turned upward in its growth, and sometimes nearly resembles fur.

"Let us now particularly consider the shapes and sizes of these escutcheons. There is one general shape to which they conform, and that is that they are wider below than above, and at or near the top of the udder they narrow in abruptly; some continue up as far as the vulva, and even above it, and others but a little distance above the udder. The size and shape of this upper part of the escutcheon is of less importance than that of the lower part, but both must be considered—the larger the escutcheon the better. All great milkers have very large escutcheons. In large ones the upturned growth often begins on the belly, in front of the udder, extends along between the teats and up the back part of the udder, over the whole width. Indeed, the udder is not wide enough for it, and it encroaches on the thighs, where we may find the hair having an upward growth on them, inside next the udder, beginning not far above the hock joints, and running up as high as the wide part of the escutcheon extends up the thighs, and which often terminates with corresponding curls in the hair at the outlines, and the higher up and wider these are apart the better. Though the extension of the escutcheon to the front part of the udder on the belly has been mentioned, that is not a matter of practical interest in ordinary cases. All that needs to be studied is plain to be seen by standing behind the cow. When the escutcheon is small, it does not reach the thighs, and often does not cover the whole of the back part of the udder. These differences in size can be distinguished at the first lesson taken in the cow-yard, and when that has been done, the next thing is to consider their shapes. A good escutcheon is symmetrical. The feathers on the two thighs are at equal distance from the middle line of the body, and extend up to equal heights on the back parts of the thighs. A broad and high escutcheon, (speaking now only of the lower broad part of it,) that is *alike on both sides*, certainly indicates a superior milker. There is nearly always another sign accompanying such an escutcheon, and that is one or two *ovals* just above the hind teats, on which a *fine* coat of hair grows downward. These may be large or small, may be one or two, and may be alike in size, or unlike, but they are always good signs. Two are better than one, and the larger and more uniform they are the better; they are almost always present on large and symmetrical escutcheons. No escutcheon is ever first class if it has not one or both, and one, at least, of good size. What constitutes 'good size' will be better learned by a few observations than can be taught by inches, and I want to leave something to the ingenuity of the learner, to make the study interesting.

"Now, let us consider the shape and size of that part of the escutcheon which I have spoken of as the upper part; that is, the narrow portion that has its base on the top of the lower and wider portions, and runs up toward the vulva. Sometimes, though very rarely, this does not exist at all. Sometimes it is broad, and extends all the way up, with perfect symmetry. Sometimes it terminates in a curved line, at a greater or less distance up; and, indeed, it may be seen of almost any shape. As a sign of excellence, the larger and more symmetrical it is, the better—but a good *lower* part of the escutcheon is the main thing, and that, as a sign, can hardly be vitiated by any imperfection of the upper part. When the lower part is very good, there is usually uniformity in the part. A poor escutcheon is one that is small, or that is imperfect in form."

The Progress of His System.

With his mind keenly alive to the pursuit of his investigations, he soon perceived the difference in the shape of these quirls or marks in the hair. We can imagine how, when he saw any cow with the same escutcheon as his own had, he would eagerly and closely question the owner, and then make his comparisons and deductions. Then, again, when he would see variations from his cow's escutcheon, whether larger or smaller, though of similar shape, how he would study them over! When he would ask of the owner such questions, directed by his knowledge of the cow's marks, the owner would stare, and think how the lad could know so well of *his* cow. And then his secret exultation when the answers showed him that he had judged aright! We can imagine this young enthusiast going on, from step to step, filling up his leisure with his acquisitions of his new theory, which was becoming fact, and growing into a system.

From his first step of discovering the dandruff, its scarcity or abundance, to his noticing the great diversity existing among cows as to the shape of the bearded ears or quirls, and being convinced these shapes were the signs by which to distinguish cows, and then to make sure that the same mark might always be relied upon as a positive sign of the same perfection or defect; were all steps in the discovery that engrossed his whole mind. He gave up his trade, traveled about, visiting cattle markets, fairs, and stables. Conversing and cross-questioning all whom he could; fixing the results in his mind, and getting the classification shaped out. He talked with farmers, dealers, and veterinary men, ascertained their modes of judging of the points of an animal, and found they were all by their own favorite signs and marks. One looked to the udder, the horns, the hide, or the shape; others to the hair, the veins, or something else; but none judged by the signs which he had found out. All were uncertain. The most the best judges could do would be to guess rightly, perhaps, three times out of five, but none could tell how long a cow would milk. Perfecting his judgment he would visit the same places and the same cows several times in a year, to see how nature was operating upon the animals, and their changes of character in the different periods of gestation, their treatment and food.

Of course, he soon began to put his theories to practical value, and he dealt in cattle on his own account. This brought before him cattle from Holland, Switzerland, Brittany, and other countries. This improved his opportunities by proving to him that, no matter what country gave them birth, all individuals possessing the same marks belonged to the same class and the same orders; in short, that nature acted through uniform laws.

Imperfections and Tufts.

Variations would arise, from crossing two animals with different escutcheons, from some defect in marking at the birth, from lack of development, or from those freaks that nature sometimes plays. They always prove stumbling-blocks in forming the judgment on some animals, and furnish texts to the opponents of the system.

As Guenon continued his examinations, he found that his classes did not afford a place for all animals, or rather that there were occasionally to be found cows whose escutcheons while apparently belonging to one of these classes, had at the same time, certain distinguished features which he styled imperfect escutcheons. These Mr. Hazard, the secretary of the commission, described as follows:

"The perfect escutcheon of each Class is the one which is in Order No. 1. All variations from this are rated lower in the scale; these variations may consist of a smaller size, therefore, the escutcheon would not be so broad or high upon the thighs, nor so broad upon the vertical portion; they may consist of the lack of ovals, which would place them below the first order; they may consist of blemishes, which are tufts of hair growing alongside of the vulva, or below it; or they may consist of strongly marked imperfections, which may be cuts or slices taken out of the escutcheon; or, coarse, harsh, wiry hair on the back and upper part of the udder. Finally, they may be so decided as to place the animal among the bastards.

Of the tufts, Guenon says all tufts encroaching on the escutcheon diminish its value, except the oval ones on the udder; that is to say, they indicate a diminished aptitude for yielding milk. The size and location of these tufts make the animals descend one or more orders in the classification. It is, therefore, important to attend to all the patches of descending hairs which lessen the size of the escutcheon, whether these occur in the middle of it or form indentations on the sides. These indentations, partly concealed by the folds of the skin, are sometimes perceived with difficulty. Many cows, which at first glance appear to be well-marked, on close examination display their deficiencies, and want of this scrutiny often causes mistakes in estimating the value of cows, and thus the system suffers.

Guenon says the cause of the defects, as exhibited by the tufts on the thighs, is that the veins situated beneath, on either side of the belly, have a peculiarity; that they are contracted, and there is a small opening for it where it pierces the abdominal muscles.

Sometimes there is an intermingling of two forms of escutcheons. This depends upon the crossing between a cow of one class and a bull of another. This is one of the difficulties to be encountered in precisely estimating the value of the animal.

Guenon classified the seven tufts, into two kinds: Those on which the hair ascends, and those on which it descends. Those with ascending hairs are simply traces which encroach on the descending hair outside the escutcheon, either on one side or beneath the vulva. Those with the descending hair are on the escutcheon, and are five in number.

1. *Epi ovale*, oval tuft. These are situated on the udder, like those on class one, two, three, four, order first. They are good signs, if of descending fine hair, small, and regular. They are mostly seen on only the best cows, though occasionally to be met with in some of the lower orders.

2. *Epi fessard*, ischiatic tuft. These are found on the vertical escutcheon on one or both sides of the vulva, as in class four, five, orders two, three, four; and very conspicuously in the bastards of class three, four, five, six. They are of ascending hair, and never seen in first class cows, but in most others to a limited extent.

3. *Epi babin*, lip-shaped tuft. This is only seen as a sign of deterioration in the two first classes; it is made by descending hairs, and is a defect for milking qualities. It is like a string hanging over the top of the vulva, and making its outline a little below it on each side. It is seldom seen

4. *Epi vulvé*, vulvan tuft. This is also a deteriorating sign; is a tuft of descending hair directly under the vulva, as in class one, orders three and four.

5. *Epi batard*, perinæal tuft. This is always a bad mark, as it exists on otherwise good marked cows, and indicates a diminution of milk, as soon as the cow becomes pregnant. It is seen on class one, bastard. A cow is to be looked upon with suspicion that has this mark largely developed.

6. *Epi cuissard*, thigh tufts. These are diminutions of the escutcheon by encroachment of descending hair, and denote a diminishing of the quantity of milk, proportionate to their extent. See class one and two, order four.

7. *Epi jonctif*, mesian tuft. The mesian or dart-like tuft, with soft silky ascending hair, is rarely seen, and only in those classes in which the escutcheon does not ascend to the vulva. It is like a V hanging beneath the vulva, and is not fully represented in the plates, though class ten, order two, shows it somewhat."

In these observations among cows, not only during their work as members of the commission, but also in preceeding examinations, Messrs. Blight, Harvey, and Hazard noticed a series of marks, which they have denominated *thigh ovals*. The plate showing the escutcheon of Mr. Hazard's Jersey cow furnishes one of the best illustration of these marks yet met with by the commission. Where the vertical escutcheon joins and widens out into the thigh escutcheon, there is usually a dip of a curved shape more or less in extent. In the plate above alluded to these thigh ovals descend nearly to the base of the udder. In their careful examination of more than two hundred cows, the commission always found these marks only on good cows.

In his examinations Guenon found cows of apparently each class with certain variations in their markings which distinguished them and prevented their incorporation into any class, and, yet the similarity gives them a claim in their particular class. In all cases he claims to have noted that cows thus marked would milk as well as other members of their class, until they were got with calf, but as soon as this was accomplished, the quantity of milk fell off rapidly. The commission claim it is this style of marking which is most likely to deceive the superficial or amateur investigators, and that these have caused the assertion that a poor cow may be well marked, when in reality, if properly understood, she was *not* well well marked. This class of cows Guenon styled *Bastards*, and he practically assigned to them a distinctive or seventh *order* in each class.

In 1822, Guenon seems to have first reduced his system to a classified basis, and from that time until 1828 he appears to have given it much of his time and attention. Having, as he deemed, sufficiently arranged and tested his system, he, in 1828, applied to the academy of Bordeaux for a public test of the correctness of his mode of judging of cows and their milking value.

The following, from the proceedings of the academy, shows that Guenon did not make his system common property. The minutes of the academy, under date of June 3, 1828, contains the following record: "Mr. Francis Guenon, of Libourne, possessor of a method which he deems infallible for judging, by mere visual examination, of the goodness of milch cows, and the quantity of milk which each can yield, has solicited the Academy to cause the efficaciousness of this method to be tested by repeated experiments. The case presented by this request was one of a secret method

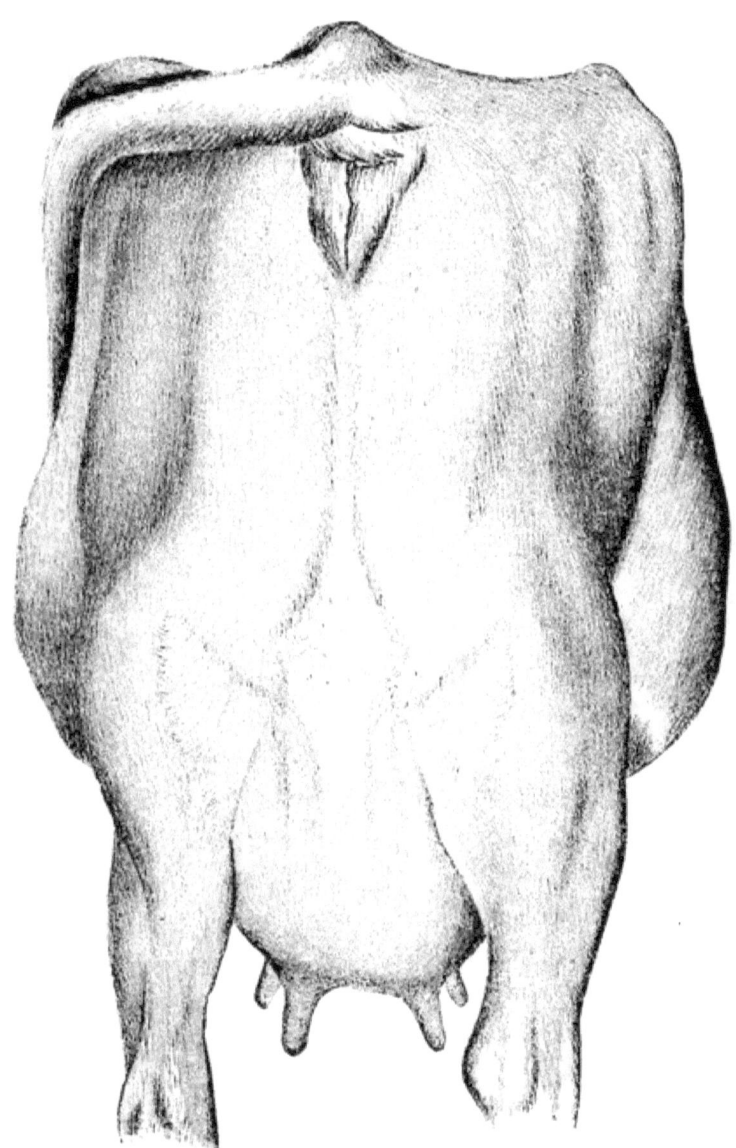

Escutcheon of ROSIE.
Thoroughbred Jersey Cow, belonging to Willis P. Hazard.

of judging, which the possessor was not willing to reveal. On the other hand, it seemed difficult to admit that the external sign, whatever it might be, by which Mr. Guenon judges, could always bear a proportional relation to the quantity of milk yielded by a cow. Nevertheless, the academy deemed it proper to appoint a committee charged with making the examination. Trials have been made with care, and under precautions necessary for precluding all collusion. The cows used for the purpose belonged to three different herds, and amounted to thirty in number, and the result has been to establish, to the satisfaction of the committee, that Mr. Guenon really possesses great sagacity in this line. So long, however, as his method shall be kept secret, it cannot be judged of, nor rewarded by, the academy. Governed by these considerations, the academy, having ascertained from Mr. Guenon that he is willing to submit to every test that may be proposed, and to disclose his secret, upon receiving a just indemnity, has referred him to the prefect, and has engaged to recommend him to the favorable notice of that magistrate, who is ever disposed to promote all that tends to improve it."

From 1822 to 1827, it would seem that Guenon perfected and studied his system, but it does not seem to have come promptly before the public, until the agricultural society of Bordeaux took upon itself a careful investigation of the whole system. From the detailed report of this committee, appointed by this society to test the knowledge of Guenon, we take the following as illustrating, not only the results reached by them, but also the manner of conducting the examination:

"Every cow subjected to examination was separated from the rest. What Mr. Guenon had to say in regard to her was taken down in writing by one of the committee; and immediately after, the proprietor, who had kept at a distance, was interrogated, and such questions put to him as would tend to confirm or disprove the judgment pronounced by Mr. Guenon. In this way we have examined, in a most careful manner—note being taken of every fact and every observation made by any one present—upward of sixty cows and heifers, and we are bound to declare that every statement made by Mr. Guenon, with respect to each of them, whether it regarded the quantity of milk, or the time during which the cow continued to give milk after being got with calf, or finally, the quality of the milk as being more or less creamy or serous, were confirmed, and its accuracy established. The only discrepancies which occurred, were some slight differences in regard to the *quantity* of milk, but these we afterward fully satisfied ourselves were caused entirely by the food of the animal being more or less abundant.

"The result of this first test seems conclusive, but they acquire new force from those of a second trial in which the method was subjected to another test through M. Guenon and his brother. Your committee, availing themselves of the presence of the latter, caused the same cows to be examined by the two brothers, but separately, so that after a cow had been inspected, and her qualities as indicated by the signs in question had been pronounced upon by one of the brothers, he was made to withdraw; then the other brother, who had been kept aloof, was called up, and desired to state the qualities of the same animal. This mode of proceeding could not fail to give rise to difference, to contradiction even, between the judgments of the two brothers, unless their method was a positive and sure one. Well, gentlemen, we must say it, this last test was absolutely decisive. Not only did the judgment of the two brothers accord perfectly together, but they were in perfect accordance also with all that was said by the proprietors in regard to the qualities, good or bad, of every animal subject to this examination."

On the 26th of May, 1837, a similar test was made by the agricultural society of Aurillac, whose committee, in their report, use the following language:

"Each cow was examined separately by M. Guenon, who wrote his notes upon her, and delivered the paper closed to one of us. Immediately after, another member of the committee questioned the owner of the cow, or the person in charge of her, in regard to her daily yield of milk, its quality, and the time during which she continued to give milk after being got with calf. The answers were taken down in writing, and then compared with the notes written by M. Guenon. They were generally found to accord, and proved to the satisfaction of your committee and of every one present, all of whom attended with lively interest to these proceedings, that M. Guenon possesses great sagacity in judging of cattle, and that his method rests upon a sure foundation."

The Bordeaux committee added: "To the proprietors and to the lookers-on, all this was very surprising for the examinations were as quickly made as the results were certain. As to ourselves to whom the method was no longer a secret, it was with renewed interest and astonishment that we viewed the accuracy of the results. *This system we do not fear to say is infallible.* We only regretted the whole society was not present."

The committee further reported that Mr. Guenon had, after more than twenty years observations and researches, discovered certain natural and positive signs that were proof against all error, while the writers and professors who have particularly occupied themselves with the bovine race, can only indicate some vague signs for judging of the fitness of cows for secreting milk. That this method is valuable, whether it tells the yield of milk only, or indicates the improvement of breeds, which are liable to deterioration from mismanagement in crossing, and that it is applicable not to full-grown animals alone, but also to calves at as early an age as three months. Thus it affords a sure means of forming a judgment of full-grown animals, about which we might be misled on account of their form and their parentage, and secures the improvement of herds by enabling us to dispose of those calves which will not repay the cost of rearing them. We shall thus no longer rear calves at great expense for two or three years that should have been consigned to the butcher, nor sell calves that would pay best to rear. If this system is pursued, only cows and bulls of best quality will be kept, and in very few years how great will be the improvement of our herds, and largely increased the cheapest and best of all foods, milk, and the production of butter and cheese.

The committee of the Agricultural Society of Bordeaux, therefore, decreed Mr. Guenon a gold medal, made him a member of the society, ordered fifty copies of his work, and distributed one thousand copies of their full report among all the agricultural societies of France.

The next public test Mr. Guenon submitted his system to, was that by the Agricultural Society of Aurillac, and that society reported that Mr. Guenon examined the herd of their president, of one hundred cows, from which were selected designedly, the best, the moderately good, and the most indifferent of the establishment. Upon each, Mr. Guenon pronounced with precision, and his decisions corresponded almost invariably with the statements of the persons in charge. The only variations were very slight ones, in regard to the quantity given. But this herd was fed unusually high, and Guenon was totally unaccustomed to the usages of the country in feeding cattle, and this caused him to pronounce the yield a little less than it really was. A proof of his system, for he declares the yield will vary according to the feed and management, which all observant farmers know to be the case. Mr. Guenon examined some of the cows a second time, and also the calves, and those calves he assigned to the first orders the cowherds said were from their best cows, that gave a great deal of milk.

The notes of his reëxaminations corresponded exactly with his first statements. The committee therefore awarded Mr. Guenon a gold medal, made him a corresponding member, subscribed for twenty-five copies of his book for each of the sub-societies, and distributed their report through all the agricultural channels of France.

With these testimonials, the highest that could be procured in France, Mr. Guenon went on with the publication of his book, which had a wide circulation in every department of France. And he was finally granted a pension for life of three thousand francs a year by the French government, after the National Assembly's committee on agriculture had given the system a thorough test. In the presence of fifty of the most eminent agriculturists, M. Guenon made his examinations, and judged correctly of all but one of the quantity, of all but one of the time, and of all of the quality; and the committee reported the results were altogether conclusive, and that his discovery had reached to the dignity of a science. They also declared the daily production of milk in France might be increased by several millions of pints daily, and that the abundance and quality of milk in the dams must contribute largely to the improvement of the progeny. They voted him the pension, and invited him to deliver lectures in the different veterinary, agricultural, and normal schools of the kingdom, and before the different agricultural societies, as "the speediest and best means of spreading the knowledge of this discovery," and " to repair the time lost in ridicule, doubt, or indifference—the inevitable preface to all undertakings beneficial to humanity."

In the foregoing account of Mr. Francis Guenon, it will be seen that, by his indomitable perseverance in perfecting his system or method, he raised himself from the ranks of a poor gardener's boy to the position of a great benefactor, and was presented with various medals and decorations, and a large sum of money voted to him. Surely, such a brilliant position must have been won entirely by merit, for he had neither means nor influence to advance him into notice.

Of the Ovals.

The ovals on the udder are spoken of by Guenon, and our experience is that they are always indicative of a good yield; particularly, when they are uniform in size and position, and of fine, soft hair, descending on the udder. But there is another set of marks, which the Pennsylvania Guenon Commission have denominated thigh ovals, which are an invariable indication of a good cow, particularly when she is otherwise well-marked. Of these, Guenon does not speak. Eusebius H. Townsend and Chalkey Harvey were the first to call attention to them, and Charles L. Sharpless has written of them. Our own cow, which took the premium over all the Jersey cows, at the fall exhibition, in 1878, of the Chester County Agricultural Society, has them most extraordinarily developed. As she is a very thorough example of this marking, we have had the likeness made of her escutcheon, and request the reader's attention to it.

Of the Bastards.

Guenon denominates those cows which give milk, much or little, so long as they are not got with calf; but, when impregnated, begin to fall off in their milk. The term he uses is *batard*, which means, in English, bastard, spurious, of a mixed breed, mongrel. We should have preferred to call them spurious cows, as the term bastard does not exactly express the meaning we apply to that word; but, as it has before been translated bastard, and is so known by many, we retain it.

The bastards are often the best looking cows; have finely developed escutcheons, and many give a great deal of milk, some poor quality and some rich; but, as soon as they are pregnant, they go dry very soon, or fall off rapidly in their milk, while others give very little milk at all. From their fine show, they deceive a great many, and Guenon cautions buyers, as the most skillful will make mistakes. He has, however, given a series of drawings, by which they can generally be discovered.

These bastards mostly conceive well, and the first time they are put to the bull, they vary in the quality of milk they give like other cows. The flow of milk is at its height during the first eight days after calving, though of bad quality. It then diminishes a little, and keeps on at about the same yield until she conceives again, when it diminishes again, more or less rapidly.

To discover a bastard, consult the engravings which are given to each class. To the first class, the Flanders, there are two kinds. The first, which is the most common, has on each edge of the vertical escutcheon, a feathery appearance, and where this is strongly marked by the down—and up-growing hairs meeting, and they interlock and stand out from the skin, and, besides, are harsh and wiry, and generally shiny, glistening, and looking of lighter color, *beware of them*. The harsher, coarser they are, the shorter time will the cow milk after getting with calf. The second kind of bastards among the Flanders will have an oval on the *vertical* escutcheon, generally near the middle part, of about two to three inches in length, by one and a half to two inches wide, on which will be found coarse wiry hair, and the harsher it is, and the larger the oval is, the sooner the cow will cease to milk. It may often be discovered by the glistening appearance of the hair on it.

On all the other classes, the bastard marks consist of two oval patches of hair, one on each side of the vulva; and the larger they are, the more pointed in shape, and the coarser and more wiry the hair on them, the sooner the cow will cease to milk.

The importance of learning the bastard marks is very great, as the buyer can safely avoid them, and leave them to those less skilled. While he may buy the less showy looking cow for much less money, and get a better animal than the unskilled man will obtain even for the higher price.

All animals are more readily judged correctly, and the system can be learned more easily, in summer than in winter, both on old and young; for then the winter coat of hair is off, and the hair is shorter, and the escutcheon is more easily perceived. The skin, also, is more natural and soft, and the hair is usually not so harsh to the feel; and the cows are cleaner, and all marks or blemishes more quickly seen.

How to Apply the System Practically.

We will now proceed to apply the foregoing rules and hints practically. In doing so, we may repeat some that has been before said, but it will only impress it the stronger in the mind of the learner.

This classification embraced all the kinds of cows known to Guenon, each individual escutcheon corresponding with one of the orders of those classes. The *class*, the *order* and the *size* of an animal indicate her yield of milk, and this will always be found to correspond with her escutcheon. Every cow has an escutcheon which can be recognized, and according as it is free from blemish or imperfection, just in that degree does she approach perfection in her class.

Guenon, in the last edition of his work, has altered and simplified his classification somewhat, for he divided it into ten CLASSES, and six ORDERS to each class. He maintained his three grades of size. But our experience shows that the cows in this country do not vary so much in size as they probably do in France, for there they have the little Brittany cow, which is very small but good, and, of course, they have also cows as large as our Durhams or the Holsteins. Only this, bear in mind, that cows, as a general rule, all other things being equal, will vary in their yield somewhat according to their size; and in judging cows apply that rule, for it is part of Guenon's system, and they will vary in the quality according to the breed. Well, then, for practical purposes, we need only study sixty *escutcheons*, that is ten different shapes called *Classes*, and six grades to each of those shapes, more or less perfect, which are called *Orders*. To these must be added ten more for a *Bastard* to each class. And it is really necessary to study perfectly only the first four orders of each class and the *Bastard* marks, as it is not worth while to purchase or pay much attention to any cows lower in the scale than the fourth order of any class. And to simplify it still more, you will notice the thigh escutcheons of the first orders have all nearly the same shovel shape, so that by remembering this you need only study the vertical portions to readily place the animals in their proper class.

The Escutcheon.

The escutcheon was so-called, we presume, from its similarity to the shape of a shield or escutcheon, and on a first-class cow it will be very like it, and some-what like a round-pointed shovel. On this escutcheon, the hair will generally be of a different color from that bordering it, most generally rather darker, always shorter, and more nearly resembling fur. This difference in color is produced by the UP-growing hair contrasting with the DOWN-growing surrounding it. The hair of the escutcheon should be short, soft, and fine; and the skin very soft, like a kid glove, thin, and oleaginous. And if the cow gives good rich milk, this skin will be of a rich, golden, or nankeen hue. Often where you handle a skin of this character the hands will feel oily, and soiled with rich dandruff.

The Shape of the Escutcheon.

The escutcheon varies in shape, and Guenon named his ten classes from their shapes.

The first class, he called Flandrine or Flanders, because it is the best, and he named it from the best cows he knew, those from Flanders, or the Flemish breed, and they had more of this shaped escutcheon than any other breed; a quiet but sure proof of the truth of his system.

The second class he called Flandrine à gauche, because although it had the Flanders shape, it was on the *left* flank, he called it therefore the Left Flanders.

The third class are the Lisière, or The Selvage, from its appearance to a selvage, or binding of a piece of cloth.

The fourth class are the Courbe-Ligne, or the Curveline, because their escutcheon is lozenge-shaped, formed by a curved line which sides to the right and left, and rises to about five or six centimeters from the vulva.

The fifth class he denominated Bicorne, or the Bicorn cow, because the upper part of this escutcheon forks in two horns.

The sixth class, Double-Lisière, or Double Selvage, has an entirely arbitrary name, and it is an odd freak of nature.

The seventh class is called Poitevine, or Demijohn, from a fancied resemblance to some kinds of demijohns.

The eighth class is Equerrine, or Square-Escutcheon, as it is square at the upward part.

The ninth class is the Limousine, as it was on a cow from that Province that Guenon first saw this shaped escutcheon.

The tenth class is called Carrésine, or Horizontal, because the upward part of the escutcheon is cut off squarely by a horizontal line.

To each of the above ten CLASSES, Guenon has placed six ORDERS, which are variations of the escutcheon, formed by a reduced size and by various imperfections. If the reader will remember always, that the first class is better than the second class, and the second class better than the third class, and so on down the scale to the end of the classes, he will have gained the first step in acquiring the system. Then the next point to remember is similar, that is, that the first *order* of every *class* is better than the second order of that class, and so on down the scale of the orders, until the sixth. Then he must learn the different shapes; first, the characteristic shape of each class, as represented by the first order of that class, and connect with this, in his mind, the number of quarts a first-class cow, in good feed and condition, should give, as represented by that escutcheon, in her full flow of milk. Then he can next learn the variations in size and shape from this pattern escutcheon, and that will enable him to tell which order of her class to put her in, and that will then inform him what quantity of milk she will give, and how long she will give it when with calf. And we repeat here, it is necessary only to acquire the knowledge of the first three or four orders of each of the ten classes, as if the cow examined does not come within those orders, she is not worth examining further nor keeping longer, nor certainly worth purchasing. Then the learner must next acquire a knowledge of the distinguishing marks which point out a Bastard cow, for an account of which marks, see under that head.

Now all of this knowledge must, to put it into profit practically, be supplemented by the careful examination of the hair and the skin, of the escutcheon, and the udder: of the hair, whether it is short, fine, soft, and furry; of the skin, whether it is soft and close-grained like a kid glove, thin, oleaginous, and yellow or golden. For if the hair is harsh, and long, particularly on the back part of the udder, it will shorten the time of giving milk, and indicates a poorer quality. The more oily or greasy to the feeling the skin of the udder and the perineum is, the more it indicates good quality and richness of milk, for the oil or fat is there, showing it is in the nature of that animal to give butyraceous milk. So with the color of the skin, if it is golden it is indicative of rich milk, and the majority think it will make a finer colored butter. There is one point more in judging by the escutcheon, and that is its size and position, and the general rule is, the higher up it is on the thighs, and the broader it is on the thighs, together with the higher and broader it is on the perineum, even up to the vulva, then the better it is. Then remember the escutcheon has two principal parts, called the thigh escutcheon and the vertical escutcheon; the thigh escutcheon extends over the udder and the thighs; and the vertical is over the perineum or that part of the posterior which extends from the udder up to the tail and above the vulva.

If the thigh escutcheon is high and broad, therefore very large, and extends far outward on to the thighs, it indicates a large flow of milk. If the vertical or upper part is broad and smooth, it indicates a prolonged flow of milk.

If the thigh or lower portion of the escutcheon is narrow, the flow will be proportionally small. If the vertical or upper part is narrow and irregular, it is unfavorable to a prolonged flow.

Chalkley Harvey says further of these marks: "Imperfections, that is blemishes of form, occur in considerable variety on both large and small escutcheons. They are all certain evidence of a diminished value of the cow as a milker. A small and imperfect escutcheon on a good cow, is something I have never yet seen. Any want of symmetry in the form of an escutcheon is an imperfection. The two sides should be alike. A small but perfect escutcheon may be better than a larger one that is imperfect. A very good one is both large and perfect.

"Thus far we have considered the escutcheon in reference to its form and size alone, and may now say, that the quantity of milk depends on these, but its quality is indicated by other signs, which we find to a great extent in the same place. It is too well known to require any assertion, that some cows give a large quantity of very poor milk, and others an equally large quantity of rich milk. It is equally well known that some cows give but little milk, though they yield a good quantity of butter; and I repeat, that the signs indicative of these differences of quality are found in the escutcheon, and they are easily recognized. If the skin in the escutcheon is soft and oily, and particularly if it is of a rich yellow color, (though this is more easily seen by examining the end of the tail,) suggestive of "gilt edged" butter, that cow will give good milk. In such cases we will find her hair soft and short. There may be some long hairs, too, but the undergrowth will be as mentioned, and often has almost the quality of fur. But if, on the other hand, the skin is white and dry, and the hair thin and harsh, the cow gives poor milk. If her escutcheon is large and symmetrical, she may give a large quantity of poor milk. The form and size of the escutcheon indicate *quantity*, the skin and hair indicate *quality*. These signs are true also as applied to bulls, being in such cases a proper guide in the selection of animals to breed milkers from. My own experience and observation, which has been considerable in the matter, convinces me that cows inherit their milking qualities more from their sires than from their dams; and it is probable that many who have been disappointed in heifers raised from some favorite milkers, will be disposed to agree with me. If this be true, then the Guenon method has an application that must prove valuable to those who breed cows for dairy purposes. Another interesting fact is, that we can discover all the signs on a calf, and are thereby enabled to select with much certainty those that are fit for the dairy, and to reject those that would be only a disappointment, if raised for that purpose. Of course, a very small cow, with ever so good an escutcheon, cannot be expected to give a very large quantity of milk, and might be inferior in that respect to one having a less perfect one, where the animal is of greater size. But in such cases, the small cow would give much more in proportion to the cost of keeping. In all cases, therefore, the size should be taken into account.

"There is a sign that may be mentioned here, (though it does not properly belong to the Guenon system,) which is a very certain evidence that a cow will give a large quantity of milk, though it expresses nothing in relation to quality. It is the large size of the vein running forward from the udder, on the belly, and just under the skin. This is called the milk vein, and when it is very large and crooked, and enters the abdomen through a hole that will allow the entrance of a man's finger, it is, I repeat, a sign that the cow will yield a large quantity of milk.

"The time that a cow will continue to milk after she is with calf, varies in different cases—some ceasing almost as soon as pregnant, and others

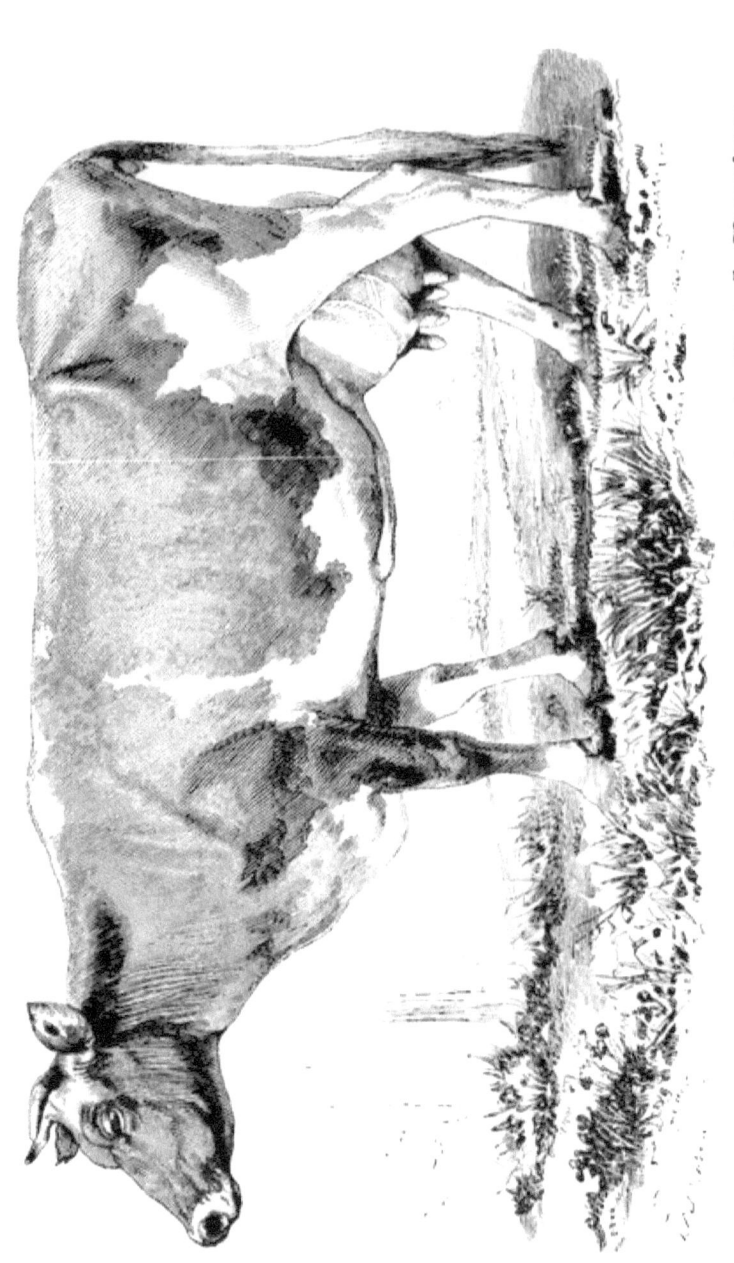

A PERFECT COW—DUCHESS—Imported Jersey, belonging to Chas. L. Sharpless.

milking up to calving. Generally the best milkers milk the longest. Hence it follows, that a good escutcheon usually indicates continued flow as well as large quantity. Those escutcheons that are not large at the base, but that run up to the vulva symmetrical all the way, and pretty wide, indicate a yield of milk up to the time of calving."

Our Mode of Judging Stock.

The beauty of the Guenon system is, that *it is an aid to all other modes of selecting stock*, and therefore, it gives a decided advantage to the person who understands it over the one who does not. For instance, let two buyers go into a herd, and let them be equal judges of stock, one of them will be very apt to buy a bastard, while the other one would very positively leave her alone, simply because the latter has a knowledge of the best and surest mode of all modes of judging stock. And this knowledge does not prevent him from using his half a dozen other modes of deciding its merits, but aids them. So, too, in selecting a bull for a propagator, the believer in Guenon will select one with a good escutcheon and a fine skin, while the other will decide almost entirely by the form. And so with calves, the one who selects calves by the Guenon marks will be pretty sure to have a dairy of productive cows, while the other will have to dispose of some unprofitable ones. The one makes money, because he is working intelligently with every light of science, while the other is only *guessing* pretty well.

We first look at a cow from the front, and see that she widens as she gets back to her hips, or is wedge-shaped. Next we look at her side, and we again see that she rises on her back and descends on the belly as she goes back to the tail, or in other words she is wedge-shaped, too, from this point of view. These two looks at her have enabled us to see that she has a feminine appearance; that her head is small and neat in proportion to her body, with a waxy small horn, a mild but large eye, a broad muzzle, and that it is well set on her neck; that she has a good chest, and large deep paunch, with large full ribs, fuller below and joined to a rather high back bone; that is to say she has not the breadth of back we look for in a beef animal. If the chine is double, it indicates a cow above the average; if the chine is single, sometimes we can lay our three fingers in three depressions in it at about the middle of it, showing that she is a loose rangy cow, and fitted for her work. Now we will look at her udder and see that it runs forward as level as possible to the belly, and that it is large, with four good-sized, well-shaped teats slightly strutting from each quarter. Now we gently approach her, and pat her to gain her confidence, and get a chance to feel her hide, her milk veins, and examine her escutcheon. If we find her skin is thin, soft, and greasy, with short fine hair, with rather a furry nature, and showing the skin yellow under it; that her udder and her perineum have soft thin skin, with very short furry hair; that her milk veins are large, zig-zag, and knotty, entering the body with good-sized holes, and particularly if this vein is double, extending and ramifying over the udder well back in prominent veins, and if the veins extend over the perineum, we may then, with great confidence, look for a large well-shaped and formed escutcheon, marked first class, order first, by an oval on each side of the back of the udder, and perhaps two thigh ovals or dips where the vertical escutcheon rises from the broad or thigh escutcheon; and just to finish and find all points corroborating, we will look on the vertical escutcheon for some spots of oily lemon colored dandruff, and at the end of her neat, lightly made tail to find some large yellow pieces of dandruff. We don't like to see it dry and brown; and as we step back from her, we just give a parting look to see that her hips are rather large, bony, somewhat drooping, that her capacious udder has room to project between her legs.

Then, we feel sure that a loose, open made cow, rather pointed, or sharp and well-defined, and the contrary of what we would look for in a flesh or beef producing animal; with a skin mellow and yellow, covered with soft, fine hair, and the nearer it comes to the quality and color of a first class Guernsey or Jersey cow, breeds which have for hundreds of years been bred for butter making, then we repeat we know she must be a good, rich milker and butter maker; for we never saw a thick, hard skin cow, with coarse, long hair, that was a good butter maker, or fit for anything but giving poor milk, *if* a strong milker.

Our preference is for a medium sized cow, one that will dress five hundred and fifty or six hundred pounds; and, as far as our observation goes, a Jersey sire, with an Ayrshire dam, is the best cross for a milk and butter cow, and the most profitable for the amount of food consumed; though a Jersey or Guernsey sire to the milking stock of Durhams, or a Holstein, or a large yielding native cow, will produce a better cow for butter than the mother was.

To get thorough practice in valueing the escutcheon, take this book in hand, and go into your dairy-yard; compare the escutcheon of each cow with her picture in this book; see what it calls for time and quantity, and then thoroughly test your cow; don't guess at it, as most farmers do; and make your own comparisons. Remember the size and class of the escutcheon will give you *the quantity and time;* the skin and hair will give you *the quality*; and always remembering the size of the cow, and of what breed she is, for they must qualify your opinion somewhat.

Opinions of the System.

A writer in the *Country Gentleman* of July 17, 1879, S. Hoxie, of Whitestown, New York, so thoroughly expresses our experience and convictions, that we are led to quote it:

"The writer has been acquainted with 'the escutcheon theory' ever since about 1850. During this time he has been a practical dairyman in central New York. At first he approached the study of the escutcheon as a doubter.

It seemed to him an absolute absurdity to claim a connection between the growing of the hair and the production of milk, two functions so entirely different.

"At first he examined the herd of cows which he helped milk every night and morning, and was surprised to meet with so many proofs of the truth of the theory. He then observed it upon other herds, and finally extended his observations to various breeds under various circumstances. He was at last compelled to come to the final conclusion that the theory, in the main, was true, but that other points and conditions of the animal must be understood in order invariably to reach a correct judgment:

"1. The breed modifies the quantity and quality of milk production. This is so manifestly true that it needs no argument. A particular order and class of escutcheon indicates a different quantity and a different quality of milk on a Jersey than it indicates on an Ayrshire cow.

"2. The condition of care and feed to which different families of the same breed have been accustomed during long periods modify milk production, and must be taken into consideration. For instance, certain families of Short-Horns have been cared for and fed through several generations with the sole view of beef production; other families have been trained to milk production. Escutcheons upon the former indicate far less quantity of milk than upon the latter. Thus some families with very fine escutcheons give very little milk. The escutcheons in such cases no doubt indicate an original capacity that a few generations of proper treatment might awaken and develop.

"3. The capacity and health of the digestive organs modifies the quantity, and we also think the quality, of milk production. Cows with large, healthy digestive organs will eat and properly digest more food, and give good return at the pail, than one with opposite conditions of the digestive organs. The former may sometimes give the larger quantity of milk, though, indeed, possessed of the poorer escutcheon.

"4. The activity of the nervous system materially affects milk production. This is often seen when the animal is unduly excited. The quiet dispositioned cow that attends to feeding, and is not disturbed by any excitement in the herd or in the surrounding fields, may have the poorer escutcheon, yet give larger quantities of milk than the extremely excitable cow, with the better escutcheon.

"Other conditions will suggest themselves to the observing and reflecting man, that materially affect the quantity and quality of milk production.

"These modifying conditions do not disturb the true theory of the escutcheon. *Other things being equal, the escutcheon is indicative of the quantity and quality of milk.* Many are misled in estimating the value of the escutcheon, because they have not the patience or the capacity to observe the varying conditions. The escutcheon is of immense practical value. It is easily seen the conditions of flesh do not change it, and animals of all ages, above three months, may be examined by it, and their milking qualities determined with a good degree of accuracy. Other things being equal, the animal with the better escutcheon will invariably make the butter maker. During nearly thirty years of observation, the writer never observed a first class cow that had a poor escutcheon. The escutcheon must be of great value to those who are breeding, and endeavouring to improve thoroughbred cattle of the various milking breeds. It offers a test that may be applied before milking age, and it may be applied to males as well as females. *Though the pedigree is ever so long, and though it contains many good ancestors, the animal should be rejected from the breeding herd, unless it has a good escutcheon.*"

"One of the Farmers," a regular correspondent of the *American Agriculturist*, writes in the number for November, 1878:

"THE VALUE OF THE GUENON MILK MIRROR.—Taken with a good udder and milk-veins, good digestive functions, and capacity for food, good health and thrift, the Guenon milk mirror is a valuable indication of both the quantity and duration of the flow of milk. This seems to be demonstrated by the experience of thousands who have given the subject careful study, and I have never yet met the man who ridiculed it, and called it "folly," who was able intelligently even to outline the prominent types. The number of calves which do well or ill as milkers, very nearly as indicated by their milk mirrors, is so large, that one of the principal practical uses to which a knowledge of the Guenon system can be applied is in selecting calves to raise, and, of course, to those who buy cows, it comes equally well in use."

THE AMERICAN ASSOCIATION OF BREEDERS OF DUTCH FRIESIAN CATTLE, composed of some of the most practical and intelligent farmers of the dairy region of central New York, have adopted a new set of rules for entry into registry in their Herd Book, wisely making the performance at the pail one of the necessary requirements. Thus, for a period of not more than twelve months from date of calving, the cow under $2\frac{1}{2}$ years of age must give 6,000 lbs. of milk; over $2\frac{1}{2}$, and under $3\frac{1}{2}$, 7,000 lbs; over $3\frac{1}{2}$, and under $4\frac{1}{2}$, 8,000 lbs; over $4\frac{1}{2}$, 9,000 lbs; also, rule 8: No animal shall be admitted to registry unless of the "milk form," or of the "combined milk and beef form," of medium or of large size, without coarseness, and if a female, having a well developed escutcheon, not below the 4th order of the 1st class, the 3d orders of the 2d, 3d, 4th, 5th, 6th, 7th, and 8th classes, the 2d order of the 9th class, or the 1st order of the 10th class of the Pennsylvania Commission. With such a record, and with such marks, no one need take the trouble to see the stock, but may safely order it, knowing exactly what they are to receive.

George E. Waring, junior, says:

"If the escutcheon teaches anything it teaches *the duration of the flow of milk*. This is its great value in connection with the Jerseys—a race of small, rich, and *persistent* milkers. It does indicate quantity, it is true, but not Dutch quantity, nor Ayrshire quantity; only *Jersey* quantity, which is quite another affair. It indicates, in at least equal degree, the continuance of the flow of milk. Indeed, this is the great value of Guenon's discovery. It is easy to judge of the *present* flow of milk in the case of any given cow, but, so far as I know, there is nothing but the escutcheon to tell us how long she will continue to milk after getting with calf. If she has a *first class* escutcheon, I think we are safe in believing that she will hold out well in her milking. If she has a very defective escutcheon, we may depend on her to fall away very rapidly when a few months gone, and to shut down entirely three or four months before calving.

From an exhaustive and admirable treatise on the Ayrshire breed, by John D. W. French, of North Andover, Mass., we make the following extracts from his remarks on the Guenon system:

"Pabst, a German farmer of large experience, with a view to simplify the method of Guenon, and render it of greater practical value, made five divisions, or classes:—

1. Very good, or extraordinary.
2. Good, or good middling.
3. Middling, and little below middling.
4. Small.
5. Very bad milkers.

"Magne, the French writer, made a still further simplification, by making four classes instead of five:—

1. The very good.
2. The good.
3. The medium.
4. The bad.

"In the first class he places cows, both parts of whose milk-mirror, the mammary and the perinean, are large, continuous, uniform, covering at least a great part of the perineum, the udder, the inner surface of the thighs, and extending more or less out upon the legs with no interruptions, or, if any, small ones, oval in form, and situated on the posterior face of the udder. Cows of this class are very rare. They give, even when small in size, from ten to fourteen quarts per day, and the largest size from eighteen to twenty-six quarts a day, and even more. They continue in milk for a long period.

"The second class is that of good cows, and to this belong the best commonly found in the market. They have the mammary part of the milk-mirror well developed, but the perinean part contracted or wholly wanting. Small cows of this class give from seven to ten or eleven quarts a day, and the largest from thirteen to seventeen quarts.

"The third class consists of middling cows. When the milk-mirror really presents only the lower or mammary part slightly developed or indented, and the perinean part contracted, narrow, and irregular, the cows are middling. Cows of this class, according to size, give from three or four to ten quarts per day.

"The fourth class is composed of bad cows. No veins are to be seen either on the perineum or the udder, while those of the belly are very slightly developed, and the mirrors are ordinarily small. These cows give only a few quarts of milk a day, and dry up a short time after calving.

"Mr. C. L. Flint, in his work on 'Milch Cows,' says:—

"These classifications, adopted by Pabst, Magne, and others, appear to be far more simple and satisfactory than the more complicated classification of Guenon. Without pretending to judge with accuracy of the quantity, the quality, or the duration which a particular size or form of the mirror will indicate, they give to Guenon the full credit of his important discovery, as a new and valuable element in forming our judgment of the milking qualities of a cow, and simply assert, with respect to the duration of the flow of milk, that the mirror that indicates the greatest quantity will also indicate the longest duration.

"My own attention was called to Guenon's method of judging cows some eight or ten years ago, and since that time I have examined many hundreds, with a view to ascertain the correctness of its main features, inquiring, at the same time, after the views and opinions of the best breeders and judges of stock, with regard to their experience and judgment of its merits; and the result of my observations has been that cows with the most perfectly developed milk-mirrors or escutcheons are, with rare exceptions, the best milkers of their breed, and that cows with small and slightly developed mirrors are, in the majority of cases, bad milkers.

"I say the best milkers of *their breed*, for I do not believe that precisely the same sized and formed milk-mirrors on a Hereford, or a Devon and an Ayrshire, or a native, will indicate anything like the same or equal milking properties. It will not do, in my opinion, to disregard the general and well-known characteristics of the breed, and rely wholly on the milk-mirror; but I think it may be safely said that, as a general rule, the best marked Hereford will turn out to be the best milker among the Herefords, all of which are poor milkers; the best marked Devon, the best among the Devons; and the best marked Ayrshire, the best among the Ayrshires; that is, it will not do to compare two animals of entirely distinct breeds by the milk-mirrors alone, without regard to the fixed habits and education, so to speak, of the breed or family to which they belong."

"In my own herd of Ayshire cows, the largest milkers have the best escutcheons, and these cows have, in most cases, transmitted these marks to their descendants. On the other hand, the cows with medium or poor escutcheons have rarely transmitted to their calves better ones; but, generally, of the same or lower class than the dams.

"BULLS.—Guenon's second and hardly less important discovery was that the bull had the same marks as the cow, only somewhat shorter and narrower. "Guenon bestows upon these marks the same name, 'milk-mirror,' which may be justified, in as far as the bull has greater influence upon the sustaining or obtaining of an abundant yield of milk, as well as the improvement of the breed."

"SOME TESTIMONY.—Mr. L. A. Hansen, of Bay St. Louis, writes, in a letter to the *Country Gentleman*:

"I served my apprenticeship for three years on a dairy farm with two hundred cows, performing all the labor appertaining to a farm, the same as one of the hired men. After this, for twenty years, I had dairies of from eighty to one hundred and seventy cows. Living in the best dairy country then known, and our butter commanding the very highest market prices in London, England, (taking the premium at a butter exhibition in London,) we considered it the best policy to buy our cows instead of raising them, and I consequently had to purchase from twenty to thirty cows every year. Having adopted the Guenon system as a helping guide in my purchases, I necessarily examined more than a hundred cows annually, besides having under daily observation my own cows and those of the neighboring dairy farms. Thus, I had continual practice through a number of years. The classifications of the professor, mentioned in my former article, were, with very rare exceptions, right. In the first two classes, they did not fail once; in the lower classes, more frequently; but as the lower classes, with their sub-division, are of no importance to the dairyman—only the two first being fit for a dairy—the study of them becomes unnecessary, and it is of little avail if they are minutely correct.

"As nothing in this world is perfect, we cannot reasonably expect the Guenon system to be without defects; but, as already stated above, the imperfection is to be looked for in that part which is immaterial for practical application. Under all circumstances, *as far as my experience goes, the Guenon theory will always remain a valuable guide in selecting milk cows.*"

"Mr. L. S. Hardin writes, in a prize essay:

"Very few, if any, modern writers upon cattle have accepted the complicated theory of Guenon, while no two of them agree as to the extent in value of the escutcheon. As a point of beauty, it should certainly be cultivated in the herd. As to its practical value for indicating the milking qualities of the cow, my experience is that a finely-developed escutcheon is rarely seen on a poor milker, while many excellent milkers have very small or no escutcheons at all. In other words, its presence is a good sign, while its absence is not necessarily a cause for distrust. Milk-veins, as an indication for milking capacity, are of about the same value as the escutcheon."

"The editor of the *Jersey Bulletin*, in commenting on this, says:

"We should be very glad to know of a cow, worthy to be called an 'excellent milker'—duration of the flow after becoming pregnant being one of the tests—which has no escutcheon at all, or a very small one. As at present advised, we don't believe she exists. Most old cow men would say that, if the escutcheon is as valuable an indication as the milk-veins, too much effort can hardly be made to extend knowledge concerning it."

"Henry Tanner, professor of agriculture, Queen's College, Birmingham, England, says, in a volume of prize essays of the Highland and Agricultural Society:

"Some attention has also been given, within a few years, to a discovery, made by Mons. Guenon, respecting 'the escutcheon,' as it is termed. Like many other persons, he was carried beyond the boundary of discretion in his speculations, and thus his valuable observations were for a time lost in the mist with which he enveloped them. Sufficient is already known of its value, at least, to lead us to the conclusion that it is worthy of more general knowledge.

"A very extended observation has proved that, other conditions being equal, the modification of form presented by the escutcheon will lead to an estimation, not only of the quantity of milk which the animal will produce, but also of the time during which the cow will keep up the supply of milk.

"Without going into detail upon this point, I may briefly state that the larger the extent of the escutcheon, the greater is the promise of milk, and also of its continuance, even after the cow is again in calf. A cow may have a small escutcheon, and yet be a good milker; but observation leads to the conclusion that, if she possessed a more fully developed escutcheon, she would have been a better milker. It may be considered a point of merit, not as deciding whether or not the cow is a good milker, but rather as an additional indication which may be taken into consideration in conjunction with other characteristic points. It is also desirable, in estimating the extent of the escutcheon, to make full allowance for the folds in the skin; otherwise, a large escutcheon may be taken for a small one. Besides the escutcheon, there are tufts of hair (epis) which have a certain degree of value when seen upon the udder of the cow."

"I presume there are many men who, although perhaps not caring a pin for an escutcheon, yet consider themselves fully capable of selecting a good milk cow. Now, although ignoring the escutcheon in their judgment, are they not apt, in selecting an ideal cow of any particular milk breed, to find a good escutcheon developed of one class or another?

"Perhaps it may be asked, if the Guenon system is a true one, why are not the Short-Horns a great milk breed, for in them we often find very large and perfect escutcheons?

"This question may be answered as follows: The Short-Horns were originally a good milking breed; but, having been made particularly a beef breed, the milking propensity or mammary system has in most families been changed or bred out. Notwithstanding this change, they may retain the escutcheon, not as a mark of quality, but as one of the characteristic marks of the breed.

"All farmers are aware that a first-class milk cow may, by injudicious feed and treatment, especially as regards milking, become a second-class animal. Now, such a system, carried out generation after generation, must certainly degenerate a milk breed, however good their marks and quality.

"Among the Short-Horns, probably the best milkers have good escutcheons; but an Ayrshire cow, with an inferior escutcheon, might be found to give more milk than a Short-Horn with a superior escutcheon, simply because one breed has been bred especially for beef, the other especially for milk.

"To show how breeding for a purpose through many generations may ultimately change qualities, let us compare the Short-Horns with the Dutch or Holsteins. The early Short-Horns, or the Teeswater breed, as it was called, was of Dutch origin, or was certainly formed by crossing the native cattle of England with stock imported from Holland. This breed was originally considered remarkable for its milking qualities.

"The Dutch breed, bred for generations for the especial purpose of milk, is to-day noted for large milkers, and among the cows may be found extraordinarily developed escutcheons.

"The following extract, from a translation from the French of Magne on milk cows, is *apropos*, as showing the difference between characteristics of breeds and qualities of the animals :

"A long, fine head, narrow towards the horns, and a slender chest are given by most writers as characteristics of a good milk cow. Now, in Flemish, Danish, Dutch, and Brittany cows the fineness of head and chest is a characteristic of these races, and not the indication of particularly developed milking qualities, being met with alike in the good and bad milkers of those races; whilst in some of the Swiss breeds, and especially in those of St. Gervais, nearly all the cows, whether good or indifferent, possess a large head and heavy chest. The farmers of Ariege, while showing us some remarkably good cows, drew our attention to their strength of chest, ampleness of the dewlap, and the volume of the head: these characteristics of race they mistake for qualities, observing them in their best cows. On the other hand, it is to be remarked that cows with fine heads are often inferior milkers. If fineness of head were a true proof of mammillary activity, would not the cows of the Durham breed be amongst the best dairy animals in the world? This characteristic cannot, therefore, be considered absolutely appreciable, as much depends on the race to which a cow may belong. It is indicative of milk only, because it is a remarkable point in those races which have produced milk cows. Thus a characteristic of race has been mistaken for a sign of particular qualities."

"If, then, we should regard the escutcheon, as well as a fine head, one of the characteristics common in the Short-Horn, it is not necessary to consider it as an indication of any particularly developed quality. Although probably the best milkers would have this sign, yet it might be regarded as a latent sign of milking qualities which had been bred out by disuse. The only fair way to judge of the value of the escutcheon in determining milking qualities, is to consider its influence in the different breeds separately, not comparing one breed with another. In judging grade cows, characteristics and blood must have a certain influence on the judgment. The general type of the animal must be considered.

"In the Ayrshire cow, we must regard the escutcheon, not as a special characteristic of the breed, but as one of the signs denoting quality.

"If the time should come when it has become so universal a sign of quality as to be considered a characteristic of the breed, then we shall have approached much nearer perfection than at present.

"Admitting that the escutcheon theory is a failure, or at least that it has failed as a test-mark of milk, have we any other mark or series of marks that have invariably given better results?

"Magne says, that in Flanders, a cow is considered a good milker, 'especially when towards the middle of the spine the apophyses (or projections) are separated or scattered so as to leave a space between of about two finger-breadths,' for the reason that, when the spine is thus formed, the haunches are better spread, and the thighs and croup larger. The other members of the body are also better developed, the basin is ampler, and the organs placed in this cavity, as well as the udder, are more voluminous.

"Now, would our dairymen consider this a more certain indication of milk, than a good escutcheon?

"Without regarding the escutcheon as an infallible sign of quality and quantity of milk, I believe it to be one of the best indications of milk, that nature has provided; but in the use of this system, we must consider:—

1. The breed.
2. The age.
3. The feed.
4. The treatment (present and past.)
5. The health.

" A good, not to say a thorough, understanding of the Guenon system, cannot be obtained by casual observation, but only by the most painstaking examination of many animals, extending over a long period of time."

Objections to the System and to the Report of the Commission.

M. Guenon in his Treatise on Milk Cows, does not give any positive reasons why the escutcheon is indicative of the yield. He rested content with the fact, that he had proved it so before many learned men, and risked his reputation upon publishing the facts. The system as far as we have been able to trace it, has always been verified by those who have *thoroughly studied it, and tested it by extended practice according to the rules of Guenon*. The principal cavilers against it, either admit they have not constantly pursued it, or show by their writings their lack of sufficient knowledge of it. The report of the Pennsylvania commission has incited several to write against the system. The principal paper produced was one read before a meeting of the State Board of Agriculture, by Eastburn Reeder, and which he had reprinted in several papers. Of this essay, it is sufficient to say, he showed he had not studied nor practiced the system thoroughly, and because he could not understand it and got befogged, he quoted a large mass of scientific matter to show the system could not be true. These attempts at argument are so quietly, but completely, set aside in the essay of Prof. D. E. Salmon, D. V. M., on Contested Dairy Questions, quoted below, that we shall not discuss them further. For we cannot any more tell *absolutely and positively* why the escutcheon reveals what it does, than we can tell why a *black* cow eating *green* grass, converts *red* blood into *white* milk, than we can tell *why* the green grass grows. In both questions at issue, we have certain facts and theories to guide our reason and judgment about them, but we know nothing *positive*, and because it is so, Mr. Reeder and Mr. Hardin won't believe it is so or can be so.

In addition to what Mons. Magne, the eminent French veterinarian, one of the most celebrated medical professors in France has written, Professor Arnold, of Rochester says, when indorsing what Magne writes:

" The size of the escutcheon is regarded as the measure of the quantity of blood supplied to the milk-producing vessels, and are evidence of their capability of elaborating milk. In the same way, the veins take up the blood, and carry it back in the milk veins which pass through the bag and along the belly, and enter the body through one or more holes, on their way to the heart. The size of these milk veins, and the holes where they enter the body, vary with the escutcheon, and like it, give evidence of the quantity of venous blood passing away, from and through the udder, and they have the same significance with reference to quantity, as the supply of arterial blood and the size of the escutcheon."

Mr. Reeder also quotes the weights of cattle given by Guenon, and triumphantly exclaims, whoever saw such small cows in this country? Guenon distinctly quotes the weights, as net dead weight, or the animal deprived of its head and horns, its hide, entrails, and feet, and gives the excellent reason for it, when he says: "If I had made the calculations for the animal on the hoof, the figures given by me would present a great difference, which would increase according to the amount of fat, sometimes to double the weight." Unfortunately, Mr. Reeder did not know enough of Guenon's facts to be aware of this clear statement, and supposed the weights were live weight.

Again, he says the commission did not examine the stock correctly. *He* would have looked at an animal, decided what escutcheon it had, or "to which class and order she belongs, and then append the figures of Guenon as the result. Any other mode of proceeding is not testing the Guenon system." Here again his lack of knowledge of the system is shown; it would be exceedingly unjust to the reputation of Guenon, as he distinctly declares the size, the age, the breed, the treatment, the season, the period of gestation, &c., shall be fully considered. It is the judgment of just such men passed upon the system, which have tended to throw any doubt upon the merit of Guenon's assertions. What would be thought of the judgment of such a person, if told by a physician to administer three things to a patient, and he gave but one, and the patient died, and he excused himself by saying, "you told me to give him medicine, and I gave it."

Then Mr. Reeder denies the value of the system for pointing out the best *feeders*. The cow which gives the most butter, and which this system will readily point out, will fatten the most rapidly when dried off; for the butyraceous particles, which go to make the butter, will be diverted from the milk and turn to fat on the animal.

Mr. Reeder objects to the report of the commission, that they "in some cases failed to classify cows," and "made incorrect classifications," and "in some cases gave different results from Guenon," and lastly "the terms employed to denote quantity, quality, and duration, are too vague, indefinite, and unsatisfactory." In all these objections, Mr. R., it will be readily seen by any practicer of the system, shows his utter ignorance of the mode of applying it.

Guenon says it is sometimes impossible to properly classify an animal, owing to the effects of a cross, or some freak of nature. In such cases they may be judged according to the escutcheon it the nearest resembles. This the commission did, but of course could not classify them.

His judgment as to "incorrect classifications" we must pass by as of no account, he not being any more capable of that than the commission.

The same may be said of "giving different results from Guenon." That is entirely a matter of judgment. Guenon says, judge of the cow by various things and then the result will approximate the amount stated to each escutcheon. Mr. Reeder says the amount set down to each escutcheon is it flexible. We prefer to follow the skill of Guenon and not the ignorance of Reeder, as it was Guenon we were appointed to test.

Finally, he objects to the terms employed to denote the significance of the escutcheon. The great difficulty of the commission was to find herds of which an accurate test of each animal had been made and kept. We believe not one farmer in one hundred thousand has such a record. Yet the commission are expected by such "infallible" advocates as Mr. R. to tell the exact character of each cow, and that record is to be set down alongside of the inaccurate record of the owner; and if they vary at all, the commission are the ones at fault. The very terms Mr. R. objects to were employed by us by special agreement with the owners, because they hesitated to say how many quarts or pounds each of their cows gave. But where there were such careful farmers as W. M. Large, M. Eastburn, J. Pyle, and M. Conard, who gave quarts, and the commission gave quarts, we would invite attention to the comparative reports as the best answer. And even in Mr. R.'s own case we ask comparison, for the reason why the commission are on most of his cows one or two quarts higher is easily accounted for, because we did not learn until after the examination that he was generally ranked by his neighbors a poor feeder, which would certainly make the difference. In the cases of such fine herds as those of S. J. Sharpless, Thomas M. Harvey, Thomas Gawthrop, and H. Preston, &c., the accounts were highly satisfactory to their owners and confirmed them in the merits of the system. For the same reasons we object to *his* test of "the system in other herds" as any proof of the merits of Guenon, for it was his interpretation of the escutcheons that is given, and it would be very unfair to judge Guenon as interpreted by one who is not an expert.

Mr. Hardin has written much against the system, but containing very little argument, and no valid objection. We will endeavor to sift out of the mass, any points made:

He thought there should be one "non-believer" on the commission, so as to "make a fair and disinterested report." What possible use he may have been is a mystery, except to cavil at what perhaps he did not understand. The commission simply put down what they interpreted the escutcheons to indicate, and the owner stated what he knew of his stock. The two accounts were brought together and compared. What more a non-believer could have done, we are at a loss to conceive.

His process of examination was laid down thus: "To take down in writing *before you see the cows*, the owners' and milkers' opinions of all the cows to be tested." "Make the owners and milkers, *out of hearing of each other*, tell you the name of the cow, her age, how much milk she gives when fresh, how much milk she gives a year, is her milk rich or poor; have you ever tested the milk by measure, or otherwise to determine the amount or its richness; what breed is she?" "Get a non-believer to make pencil sketches of each escutcheon." "The Governor to appoint two more on the committee who are not believers."

Now, having laid out this programme, he does not say what was to be done with it. The inference was to be drawn, we suppose, that the many escutcheons were to be engraved, and the public were to draw their conclusions from them and the reports given by the owners and milkers, and see how Guenon would stand the test. And what were the believing or non-believing commissioners to do? Surpervise the taking down of all this? How, at once, this shows Mr. Hardin to know little or nothing of the system! Like Mr. Reeder, he did not know that Guenon assigns many other things to be thought of to form a correct opinion! Was it more proof to be told by the owner all that any one could know about the cow, and then say that corresponds with the escutcheon? Or did it put the system to a severer test, to say to the owner, don't tell me a word, and then proceed to tell him all about a cow you never saw, simply from examining her escutcheon? In one case, you are assisted to define the escutcheon by the knowledge given you. In the other case, you define the cow's character by only the knowledge you can get from the escutcheon. No better proof can be given of Mr. Hardin's lack of practical knowledge of the system.

Another objection he makes, and repeats several times, as being a very strong one with him, is, why did not Guenon, and why do not the commissioners, go to work and buy up all the best cows and sell them at a profit, and thus get very rich. His cry is, why don't they make plenty of money out of it, if it is so valuable? Simply, because neither of them are in that business, or care to be. But Mr. Harvey, a manager of the Delaware county almshouse, in one year from taking this position, changed the cows there, and increased the yield twofold from the same number of cows, and has bought and sold all the steers and cows on his large farm for many years solely by this system, and *has* grown wealthy.

He says in another article " feeling the modesty that naturally attaches itself to benighted ignorance," he "started out in the city in search of some one who was learned on these subjects." He found "a professor in our medical institute," "one of our most learned physicians," and they proceed together to canvass Professors Magne and Arnold's theories and facts about the formation of the escutcheon. The result of two such wise heads (or of "benighted ignorance") coming together, was that neither of them ever heard of Professor Magne, and that his dictum was "opposed to all the teachings of physiology." The learned professor knowing as much about a cow as he did of physiology. And it is such stuff as this which forms the arguments of Mr. Hardin. Professor Salmon in his essay on Contested Dairy Questions effectually settles these "learned" men.

We have devoted enough space to a writer, who finds it so easy to tear down, but is never able to build up, a doubting Thomas, whose only mode of judging a cow, he says is a crumple horn, a large udder, and to test the milk every Monday for one year. What an amount of money the farmers of America would lose annually if they followed his rules, and what an amount they would save by following Guenon's rules!

The following valuble essay is from the *Country Gentleman* of August 7, 1879:

Contested Dairy Questions.

By D. E. SALMON, D. V. M.

Several of our prominent dairy writers have been lately discussing the more complicated questions of their department in a very energetic and decided, if not in a scrupulously exact manner. Now, if these questions are worth the time and space necessary for their presentation at length, they are certainly of sufficient importance to receive candid and perfectly truthful treatment; and, though these writers may not have intended to give wrong impressions, their teachings can hardly be considered, in several respects, as representing the present condition of knowledge on these points.

MAGNE'S THEORY OF THE ESCUTCHEON.—In Mr. Eastburn Reeder's essay on the escutcheon—which is a valuable paper, though marred in the above respects—there is an attempt at scientific argument in order to ridicule the accepted value of the milk-mirror; and the assumed facts on which this argument is based, are presented in such a positive manner that they will probably be accepted, without further investigation, by the majority of readers unless contested at once. The writer has hesitated to do this in the hope that it would be done by some one else; but the truth is of too much consequence to allow the matter to pass entirely without notice.

The first point to which I will call attention is the attempt to dispute Magne's opinion that the hair turns in the direction in which the arteries ramify, and that the reversed hair on the udder and adjacent parts indicates the termination of the arteries which supply the udder with blood. When these arteries are large, he holds, they extend through the udder upward and onward, ramifying on the skin beyond the udder, and giving the hair the peculiar appearance which distinguishes it from the rest of the surface. If these arteries are very small, they are not likely to extend much beyond the udder, and, hence, form a small escutcheon; consequently, a small escutcheon indicates a feeble supply of blood, and little material to make milk of.

Now how is this combatted? The first argument is that "when Mr. Hardin showed this paragraph to one of the most learned medical professors at Louisville, Kentucky, he at once wanted to know who this Magne was, and declared his name unknown in the annals of medical science." What are we to think of such a statement as that? Magne—member of the French Academy of Medicine, formerly director of the Alfort Veterinary School and professor of Lyons—unknown in the annals of medicine!

We are then asked if the arteries are not the same in all cows, and are told that we might as well expect more bones or muscles as more arteries. If Mr. Reeder will turn to Chauveau's Anatomy—one of the best authorities in the world—he will find, in general remarks on arteries, the following statement, which I translate, not having the English edition: "Arteries very often present variations in their deposition, which the surgeon should keep in mind. These variations ordinarily concern the number, the point of origin, and the volume of the vessels." And if he will go through the list of arteries, he will find examples given of each of these variations.

Again, he asks, "how is it that the ramification of the arterial circulation *causes* the hair to grow in one direction on one part of the cow's body, and in the opposite on other parts?" Not a very difficult question, if we admit that arteries have such an effect, for they certainly do not all ramify in the same direction.

In a revised edition of the essay, subsequently published, some important points were added. Here we are told that "the arteries supplying the udder with blood are called the *mammary* arteries, and their ramification *does not extend beyond the outer surface of the udder.* Further down the *aorta*, or main artery, another pair of arteries branches off, called the *femoral* arteries. These supply the muscles of the thigh, or what we know as the *rounds of beef*, with blood, and ramify upon the portion of the escutcheon lying between them. Still further down, another pair of arteries, called the *gluteal* arteries, leave the *aorta*, and are distributed through the pelvic region, and ramify upon the extreme upper portion of the escutcheon. Here we have at least three distinct systems of arteries ramifying upon the escutcheon, and *two* of them most certainly have no connection with the milk secretion whatever."

Without attempting to point out *all* the errors of this description, we will once more refer to Chauveau to settle the more important points. The reader will find in that work that the *femoral* arteries have a branch called the *pre-pubic*, which in turn has a branch called the *external pudic*, from which the *mammary* artery branches. It will also be found that the mammary artery "sends several divisions to the tissue of the udder, and is prolonged between the thighs by a perineal branch, which terminates in the inferior commissure of the vulva, after having furnished glandular and *cutaneous* divisions." Turning to the description of the gluteal arteries, we find that they ramify in the gluteal muscles, which are at a considerable distance from the perineum, and that nothing is said of their going to the last named part.

Here, then, is complete and positive refutation of these arguments—not by mere statements of my own, but by the words of a standard work, of world-wide reputation, on the anatomy of these animals. Magne's *facts* are correct, then, whether his inferences are or not. *The same artery that supplies the udder with blood supplies the skin on which the escutcheon is formed; and, more than this, the artery ramifies in the direction in which the hair of the escutcheon grows.* Is there any connection between the two for all that? Who knows? A point or two to show that such a connection is not beyond the possible may still be in place.

Erasmus Wilson, who has made a specialty of the skin and its diseases, shows that the direction of the hairs on the anterior surface of the human body is, commencing at a point near the arm-pit, downwards and slightly inwards towards the umbilicus, and that below this point the direction is upwards and inwards; so that the umbilicus "is the center of convergence of four streams," as he expresses it.

Now this disposition, complicated though it is, certainly resembles that of the arteries—the branches from the *axillary* artery passing downwards and inwards, while the *epigastric* arteries branch from the *femorals* near the groin, and have a direction upwards and inwards. On the neck, the direction of the hair is upwards and backwards; in front of the ear, it is downwards and forwards; behind the ear, it is backwards—in each case following the arterial ramifications. In addition, Tisserant and others in France, who stand high as authorities, admit that the escutcheon continues to increase in relative surface till the second or third milking—that is, till the development of the udder, and, consequently, of the vessels supplying it have reached their highest point.

In some cases, it must be confessed, the correspondence in question apparently does not exist, but rather the opposite; and as the mammary artery has substantially the same distribution with horses as with cattle, we cannot see why the former should not be as plainly marked as the latter, if the direction of the hair depends on the direction of the arteries.

But, it may be asked, in what possible manner could the one condition influence the other? It must be remembered that physiology is still a growing science, and that there are many things yet to learn, so that it is still pardonable to confess ignorance. We know, however, that the cavity in the skin surrounding the hair (hair follicle) is set in an oblique direction, as well as the hair that emerges from it; the papilla at the bottom of this cavity must also be inclined, and it is this that, in all probability, decides the direction of the hair, as the growth of this takes place by additions of cells from the surface of the papilla. Now, each papilla, or elevation, has a vascular loop, or, as some say, a minute artery and vein, and one can easily imagine how the direction of this minute artery might influence the direction of the papillary summit, and, consequently, of the hair that grows from it.

I do not say that this is the proper explanation, but I suggest it as one way in which the correspondence might be accounted for. I do say, however, that the evidence brought to bear on this point by Mr. Reeder can have no influence in deciding the question, for the reason I have given.

Dr. Henry Stewart, the noted scientific and practical farmer and writer, said lately; "I have for some time past been studying the nature of the escutcheon physiogically and anatomically." And he has "recently discovered a still more satisfactory connection between the milking capacity of a cow and the development of the escutcheon."

"The milk-vein is an important mark of the deep-milking cow. But it is not the veins, but the arteries, which supply blood to the system, either for the production of tissue or the secretion of the milk. And yet the veins are important because they bear a direct relation to the arteries, being the return channels for the blood after it has fulfilled its functions; and so the larger supply of blood conveyed by the arteries requiring a vein of large capacity to return it, this vein is an ultimate indication of the vigor of the circulation of the lacteal organs. The main artery which supplies these organs is the subcutaneous abdominal [what Mr. S. says is commonly called the milk-vein.] This important artery supplies a large part of the posterior portion of the system, furnishing blood to the genital organs and the skin covering these and the adjacent parts. The subcutaneous abdominal artery is one of the two branches of the external pudic artery in the female, the other being the mammary artery. This last is very voluminous and distributes several main branches to the mammary glands and tissue, and also by a prolongation between the thighs, supplies the inferior commissure of the vulva and gives off many smaller branches, which spread into a network among the glandular tissue and the cutaneous structure. Here is the close connection, then, between the skin of the posterior part of the cow, from the lower point of the vulva down between the thighs and around the udder, and the udder itself. The same artery supplies all this portion of the skin, furnishes the subaceous glands and the hair follicles, and the whole cutaneous structure, and the hair also with blood, and also provides for the demands of the milk-secreting organs. A vigorous circulation through a voluminous arterial system * * * gives a relatively vigorous milk secretion, and, as well, a growth of hair, which curls and forms the well-known peculiar structure of the escutcheon."

C. L. SHARPLESS ON THE ESCUTCHEON.

We extract from our book on "The Jersey, Guernsey, and Alderney Cow," some remarks on the escutcheon, by Charles L. Sharpless, of Philadelphia. We consider him one of the best judges, a most intelligent breeder, and he has paid the highest price ever given for a Jersey cow in this country. The portraits of Duchess, Rosa, Black Bess, Tiberia, and the bull, Comet of M., bear out our assertion.

"There is no point in judging a cow so little understood as the escutcheon. The conclusion of almost every one is, that her escutcheon is good, if there be a broad band of uprunning hair from the udder to the vulva, and around it—see Fig. 1. These cows, with the broad vertical escutcheon, are nearly always parallel cows; that is, with bodies long, but not large, and with the under line parallel with the back. Their thighs are thin, and the thigh escutcheon shows on the inside of the thigh, rather than on its rear.

"Next comes the wedge-shaped cow, with the body shorter, but very large, deep in the flank, and very capacious. This form does not usually exhibit the broad vertical escutcheon, running up to the vulva, but with a broader thigh may exhibit a thigh escutcheon, which is preferable to the other, thus—see Fig. 2.

"In both vertical and thigh mirrors, where the hair runs down, intruding on the udder, (as low as above the dotted lines,) as in Figs. 3 and 4, it damages the escutcheon. If you find a cow with the hair all running down, and between the thighs—that is, with no up-running hair—stamp her as a cipher for milk-yielding.

"The artist has made the udders to Figs. 1, 2, 3, 4 the same size, while in reality they will vary according to the escutcheon.

"There are times when the udder of a cow, with an escutcheon like Fig. 4, will be enlarged by non-milking, for the purpose of deception. It is always safer to judge by the escutcheon, rather than by the large size of the udder.

"The escutcheons of the best cows—those yielding the most and continuing the longest—will be found to be those which conform to Fig. 2. [Mr. S. alludes to the selvage: one of the best, and common among the best cows. H.]

"The vertical escutcheon of Fig. 1, would not injure it; but if that ornamental feature has to be at the expense of the thigh escutcheon, Fig. 2 is best as it is.

"Whenever an escutcheon is accompanied by a curl on each hind-quarter of the udder, it indicates a yield of the highest order. * * *.

"So far we have noticed only the rear escutcheon, or that which represents the two hind-quarters of the udder. The two front-quarters are just as important, and should be capacious, and run well forward under the body—see A. If the udder, in front, be concave, or cut up as in B, indicating small capacity, it represents reduced yield.

This front or level escutcheon is distinctly marked in the young heifer or bull, and can be seen by laying the animal on its back. The udder hair under the body all runs backward, commencing at the forward line of the escutcheon—see dotted lines in Figs. 6, 7, 8. This dividing line is very perceptible, from the fact that the hair in front of it all runs forward towards the head of the animal, while the escutcheon, or udder hair, all runs backward over the forward quarters of the udder, around and beyond

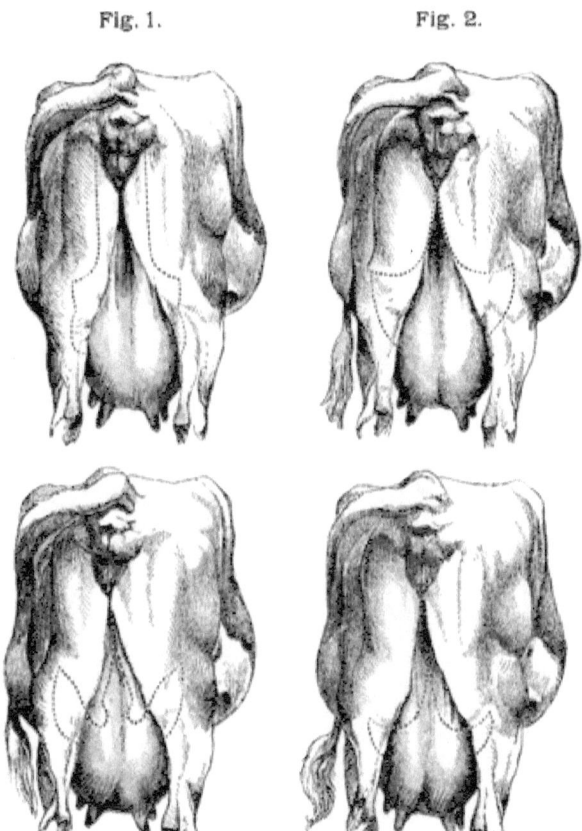

inside of the rump-bone at the setting-on of the tail. Let the teats be well apart; let them yield a full and free stream, and be large enough to fill the hand without the necessity in milking of pulling them between the thumb and forefingers. And let us ever keep in mind that the large yielder must be well fed."

Those who condemn Jersey cows as small yielders of milk and butter, should listen to the story of "Rosa" as told by her owner, C. L. Sharpless. She is five years old, is solid creamy fawn, and, combined with great volume and bone, she is neat in the head and neck, and with fine legs. Her dam was a small mouse-colored cow, and her sire's dam a small fawn-colored, neither of which would give over twelve quarts.

"We found we were making a good deal of butter, and as 'Rosa' looked superbly, we determined to test her butter quality. We fed her per day twenty pounds of hay, eight quarts of meal, and four quarts of carrots. The meal was a mixture of good wheat bran and cornmeal, in the proportion of four bushels of the former to one bushel of the latter. Her yield the first day was sixteen quarts, the second day fifteen and a half quarts, the third day sixteen quarts, and the next morning eight quarts, being in all seven milkings, or half the week. Her milk was kept separate; was skimmed after standing thirty-six hours, and made six and three fourths pounds of butter, or thirteen and a half pounds for the week.

"As you place Rosa and Duchess side by side there are some points of agreement and of difference that are of interest to notice. They are both wedge-shaped, with large body—Duchess the more bony, but Rosa with the greater rear volume, (broader hips, &c.) They both have neat heads and necks, and fine bone. Duchess is, in winter, smoke-color, with brilliant white, but not with black points. She has yellow hoofs and skin, and her udder is rich yellow. Rosa has yellow hoofs, and yellow inside her ears, but a pale skin and udder, and would be called a butter cow inferior to Duchess, and yet she has just proved herself one half pound greater. The color of it is the deepest—no coloring matter being used. This upsets the theory that a yellow skin is essential for deep-colored butter. Perhaps a safer way to put it is, that though a rich yellow skin is evidence of butter quality, yet equally good quality may come from a pale skin, provided the cow has yellow inside her ears.

"Again, as to vertical or rear escutcheons both these cows exhibit, the broad part diminishes as it rises, until, when within six to nine inches of the vulva, it is reduced to the breadth of not over an inch wide. Thus they agree in their rear escutcheons, and they agree also in udders of great capacity, these being deep and broad, and running well forward under the body.

"There is a point on which they differ. The hair on Duchess is soft and furry as a mole; that of Rosa is fairly fine, but still hair.

"So that in a word one can say soft hair, a large escutcheon, and a yellow skin are desirable, but there may be choice cows not conspicuous, for either.

"To show how we sometimes let our best animals slip, I will add that when Rosa was a heifer I was tempted to part with her for what seemed a great price—$500. In about two weeks she had a heifer calf, for which her owner was offered $150. When three years old she had a second heifer, which he sold for $180; and when four years old she had a third heifer calf, which he sold for $100. He then sold his place and all his stock, and I bought her at public sale for $375 for her beauty. Her pale skin deceived me as to her butter quality, and her, as I thought, deficient

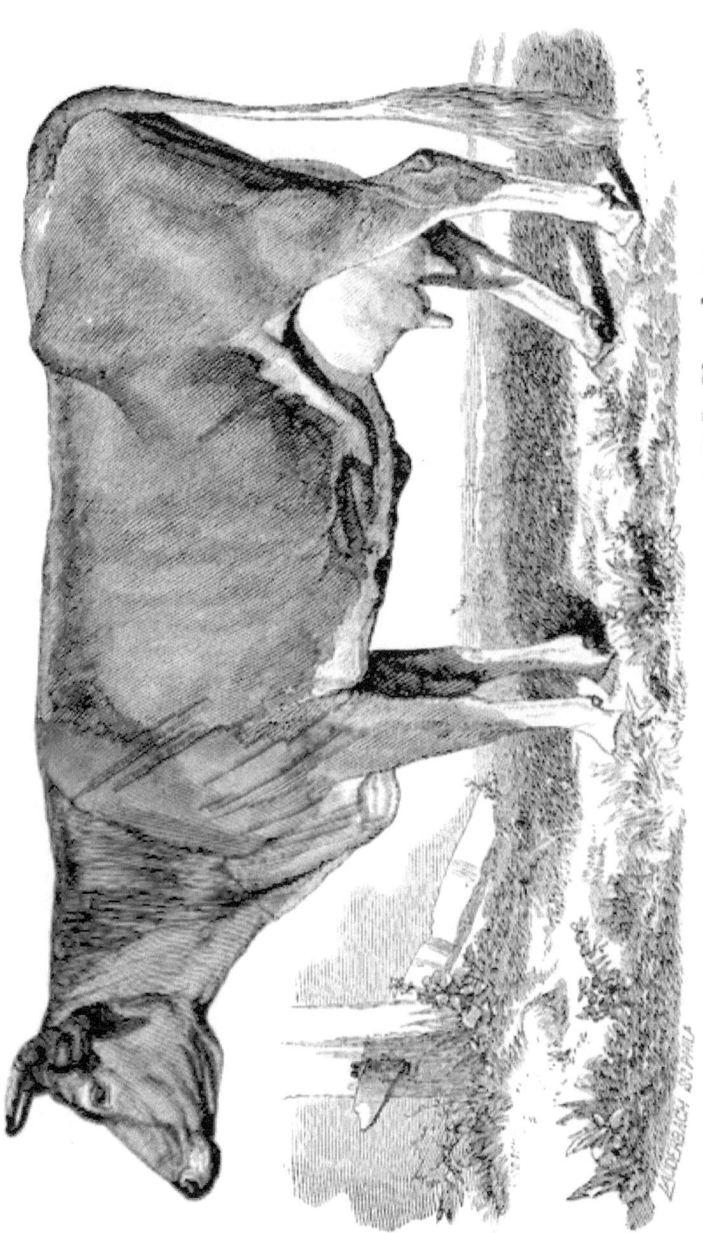

ROSA.—Imported Jersey, belonging to C. L. Sharpless.

DESCRIPTION OF THE CLASSES AND ORDERS.

Prepared by W. P. HAZARD, *Secretary of the Pennsylvania Guenon Commission.*

In the following descriptions of the ten classes, and their sub-division into six orders each, we give the quantity as stated, for a large-sized cow. Not thinking it worth while to enter so minutely into his sub-divisions of high, medium, and low cows. For instance, to class one, order one, he gives to the high cow twenty-four litres, which is about equal to our twenty-four quarts here; the litre being exactly two and one eighth wine pints. To the medium cow he gives nineteen quarts, and to the low cow, fourteen quarts, per day. The size of his high cow is five hundred and fifty to six hundred and fifty pounds, *dressed weight;* the medium, three hundred and twenty-five to four hundred and fifty pounds; and the low, one hundred and ten to two hundred and twenty-five pounds. As most of our cows will range with the high cows, we have adopted the scale suitable to the size, only the reader who practices the system must keep in mind that the larger and more developed the cow, the more she will be likely to give than the cow of smaller size.

First Class. The Flanders Cow.

Cows with this escutcheon are the most seldom found, except among the most abundant milkers. In the first order they give twenty quarts per day, in the height of their flow; that is to say, from the time they have calved until they are pregnant again. Then they diminish, little by little, until their next calving. It is best to dry them off from four to six weeks before calving, to give them a needed rest, and it improves the calf.

Cows of the first class have a soft udder, with fine hair on it, rising until it blends with similar hair growing upward on the thighs, above the hock, and widening on the thick part of the thigh, then narrowing, like in the engraving, until it reaches the vulva, and being about two inches on each side of it. The inner parts of the thigh, and the vertical mirror are usually of a yellowish or nankeen color, with dark spots on them, from which can be detached the dandruff. There are two ovals on the udder, of fine short hair.

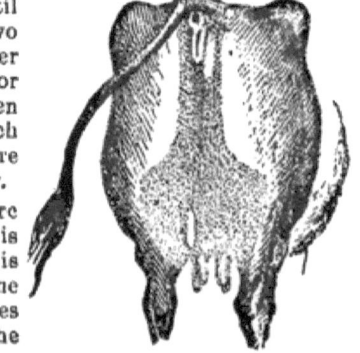

The second order of the first class are similar to the first, but the escutcheon is smaller; and on the right side of the vulva is a tuft of descending hair about two and one half inches long and one and one half inches broad, and there is but one oval on the

udder. They yield eighteen quarts of milk for a period of eight months.

The third order of the first class is still smaller, and not quite so decided in shape. It has also a semi-circular tuft below the vulva of small size, of descending hair, rather shining and of brighter color. There is either only one oval on the udder, or generally none.

Cows of the third order yield sixteen quarts, and milk for six months.

The fourth order of the first class, besides being still smaller, has narrower thigh escutcheons, and lower down; also the tuft under the vulva is quite long, about five or six inches, which sometimes make the vertical escutcheon terminate in a fork. This tuft has more lustre and is whiter than the hair around it. There is also a thigh tuft of half oval shape on the right of the escutcheon, about five inches high.

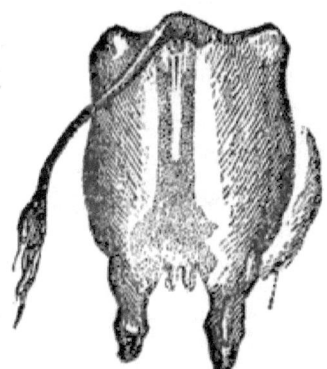

Cows of the fourth order yield twelve quarts a day, and milk five months.

The Bastard Flanders have two marks which distinguish them: 1. Some have on the vertical escutcheon an oval tuft, about the middle of it; this tuft has descending hair, is about three inches long and two inches wide, and the lustre of the hair makes it appear as if it was whiter than that around it. The larger the oval the sooner the milk will fail, and the smaller it is the longer will she milk. 2. Other Bastards of this class are distinguished by the ascending and descending hair interfering with each other on the outlines of the vertical escutcheon, looking feathery, or bristling like the beard of wheat. The skin is fine and reddish, but there is no dandruff. The larger the escutcheon, and the finer the hair, the more abundant the milk; but when the hair is coarse, long, and thin, the yield is small. Both kinds of Bastards of this class have every other appearance of the best cows. And all Bastards of the first classes have the two ovals on the udder.

Second Class. Left Flanders.

The cows of this class are very similar to those of the first class, though their yield is rather less. The vertical escutcheon is entirely to the left of the vulva, and the thigh escutcheon on the right is broader than that on the left. By comparison with the first class, these will be seen to be very similar, but in each order smaller; therefore, it will not be necessary to describe them separately, but simply to state the yield. Cows of the first order of the second class will yield eighteen quarts, and milk eight months.

The second order of the second class have the lip-shaped tuft to the left of the vulva, and have one oval on the left of the udder. Cows of this order give sixteen quarts, and milk seven months.

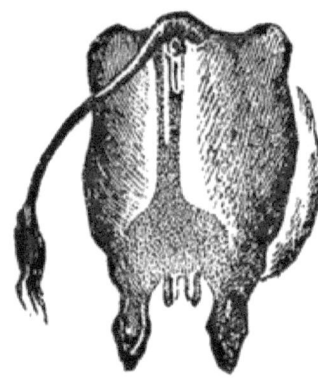

The third order has the same shaped escutcheon, but more contracted, and the lip-shaped tuft is larger and whiter. Cows of this order give fourteen quarts, and milk six months.

The fourth order has two invasions of the thigh escutcheon by the down-growing hair, a semi-oval one on the right, and a triangular one on the left. These always indicate a reduced quantity of milk.

Cows of the fourth order give ten quarts, and milk five months

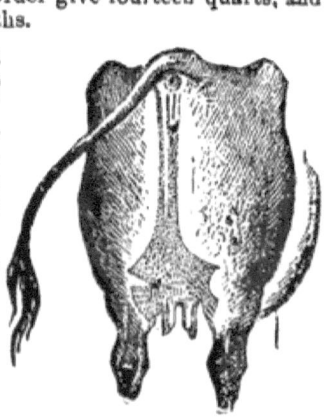

The escutcheon of the Bastard Left Flanders is known by this peculiarity. The developments are larger and more irregular on the top of the vertical escutcheon, and to the left of the vulva; on the right is the ischiatic tuft, quite large, from which the hair is diverted in an almost horizontal direction.

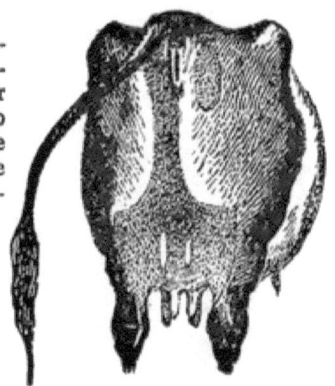

Third Class.—Selvage.

The escutcheon of this class commences above the hock, runs up on the thighs quite high, thence it descends somewhat from both sides to the vertical portion, which rises, gradually narrowing to the vulva.

The first order of the third class has an udder with soft skin, and fine downy hair, which, as well as the thighs, are of a yellow or nankeen cast of color. There are two ovals on the udder. Cows of the first order give nineteen quarts, and milk eight months, and often will milk nine months, not going dry unless made to.

The second order is similar to the first, only of reduced size; it has a tuft to the left of the vulva; and only one oval on the udder on the left side; the hair of the escutcheon is generally more glossy than that around it. Cows of the second order give seventeen quarts, and milk seven months.

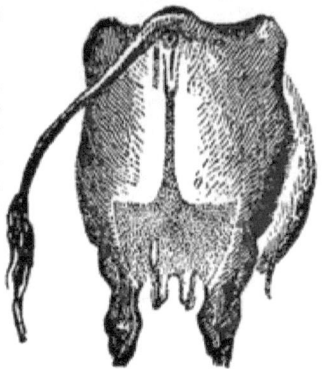

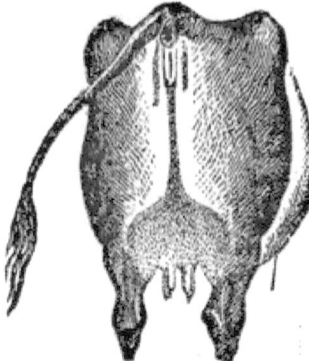

The third order escutcheon curves downward on each side of the vertical mirror, which rises narrowing to a point at the vulva; to the right and left of the vulva are tufts, the one on the left being the longest; on the left of the udder is sometimes an oval. Cows of the third order give fifteen quarts, and milk six months.

The escutcheon of the fourth order is of similar shape, but still smaller; 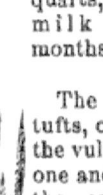 but the tuft on the left of the vulva is much longer than on the right, and there is no oval on the udder.

Cows of the fourth order give twelve quarts, and milk five months.

The bastards of the third class have two tufts, one on the right, and one on the left of the vulva, about four to five inches long, and one and one half inches wide. The smaller they are, and the finer the hair on them, the less rapid is the loss of milk. But if they are large and have coarse hair, and are pointed at each end, they prove the milk to be poor and serous, and the cow will fail rapidly.

The Fourth Class. Curveline.

The Curveline cows are very plenty, and are of a very good grade, approaching the first class. The escutcheon is broader than the last two classes, in the upper part. Their skin is of delicate texture, and nankeen shade of color on the escutcheon. The higher and broader the curved line rises toward the vulva, which it never reaches, the better it is. There are two ovals on the udder. Cows of the first order of the fourth class give 19 quarts, and milk eight months, and sometimes up to their next calf.

The second order have the same shape escutcheon, but more contracted. There is but one, and sometimes no oval on the udder. On the left of the vulva is a small tuft.

Cows of the second order give seventeen quarts, and milk seven months.

The third order has a smaller escutcheon, with two tufts by the vulva, the left longer than the right, about four inches by one inch wide. Sometimes an oval on the left side of the udder.

Cows of the third order give fifteen quarts, and milk six months.

The fourth order has a much smaller escutcheon, reaching just above the udder. The two tufts are larger alongside the vulva, and the hairs bristle to each side. On the right, the downgrowing hair intrudes somewhat upon the escutcheon.

Cows of the fourth order give twelve quarts, and milk five months.

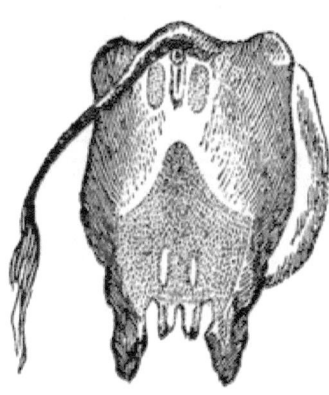

The Bastards of the fourth class have a fine appearance of escutcheon, but they are known by the tufts alongside the vulva. If they have coarse bristly hair, and of large size, say four to five inches long, and one and a half in width, they indicate a rapid loss of milk as soon as pregnant, particularly if they are pointed at each end.

The Fifth Class. The Bicorn.

The escutcheons of this class in the vertical portion end below the vulva in an indented shape, presenting the appearance of two upright horns. Their udders are of a saffron color, delicate, with fine, soft hair, and have much dandruff.

The first order has two tufts of small size along the vulva, and two ovals on the udder. They give seventeen quarts, and milk eight months.

The second order are similar to the first, only smaller escutcheons, the vulva tufts are longer, and there is but one oval on the udder, on the left. The right horn of the escutcheon is smaller than the left one.
Cows of the second order give fifteen quarts, and milk seven months.

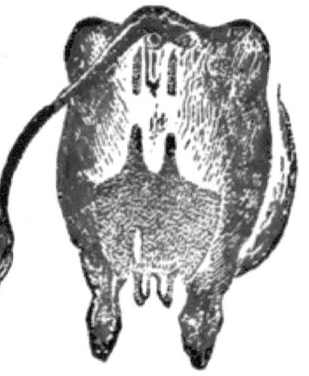

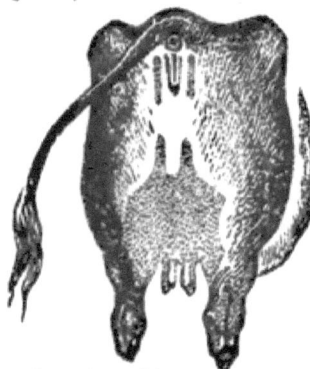

The third order have similar escutcheons to the last, but smaller, while the vulva tufts are larger, there are no ovals, and there is an invasion of the descending hair on the right side. The right is two inches shorter than the left.
Cows of the third order give thirteen quarts, and milk six months.

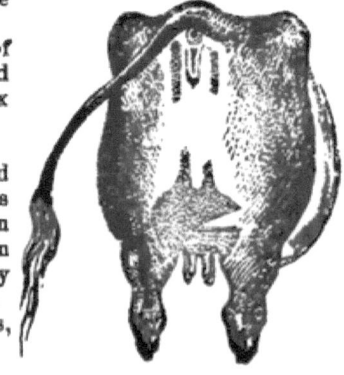

The fourth order have the same shaped escutcheon, but smaller, with two tufts alongside the vulva, larger than those on the last. On the right of the escutcheon is a triangular cut in the shape, made by encroachments of the down-growing hair.
Cows of the fourth order give ten quarts, and milk five months.

The Bastards of the fifth class have the full escutcheon of the first or second orders, but with two large tufts alongside the vulva, which, according to their size, and more or less pointed shape, and fine or coarse hair, indicate the more or less stoppage of the flow of milk.

The Sixth Class. Double Selvage.

The escutcheons of Double Selvage cows differ from those of Selvage, or the third class, in that the escutcheon is marked in its whole length by a strip of hair descending and dividing it into two equal portions. It is bordered in its whole length and at the extremity by a double line of ascending hair, which extends the escutcheon up to the vulva. Otherwise it is like the selvage escutcheon.

The first order cows have a fine udder, soft, and covered with a silky down; and its skin is yellowish or nankeen. Cows of the first order give eighteen quarts, and milk full eight months.

The second order have a similar escutcheon, but smaller, and the separating strip ends higher up. Cows of the second order give sixteen quarts, and milk seven months.

The third order have a still more reduced escutcheon, the descending strip terminating at the udder.

Cows of the third order give fourteen quarts, and milk six months.

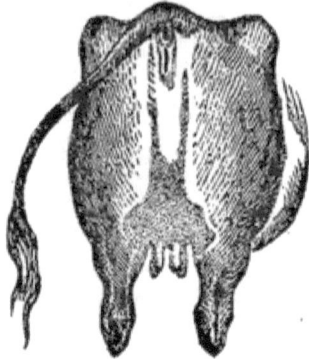

The fourth order have an escutcheon more broken in appearance, the two side lines of the selvage terminate half way to the vulva, and end off in lines of a feathery appearance, the hair is coarser and more furry,

Cows of the fourth order give ten quarts, and milk five months.

Bastards of the sixth class have the escutcheon similar to the first class, but the selvage lines terminate on each side of the vulva in tufts of coarse and bristly hair; the larger and coarser they are, the sooner the milk will fail.

The Seventh Class. Demijohn.

The first order of this class has the skin of the escutcheon of yellowish color. The udder is fine, and covered with a silky down to the inside of the thighs; and the dandruff is soft and oily to the touch. The shape is similar to the selvage somewhat, only the vertical mirror rises broader and straighter, and ends half way up to the vulva, cut square off. The broader and higher this part is, the better the escutcheon. The escutcheon is not so high up on the thighs as the previous classes. There are two ovals on the udder, and two small tufts of fine hair alongside the vulva.

Cows of the first order give seventeen quarts, and milk eight months.

The second order have the escutcheon lower down and, of course, smaller in every way. There are two tufts alongside the vulva, the left one as large as in the first order, (two and a half inches,) the right one only half as long. There is one oval on the left of the udder.

Cows of the second order give fifteen quarts, and milk seven months.

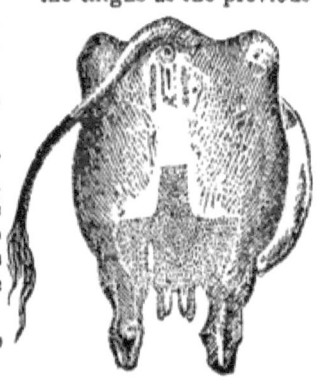

The third order escutcheon is of different shape, the lines converging downwards from the vertical mirror, which is short, and cut off square. The right side frequently has a curved line from the descending hair invading it. The vulva tufts are longer than in the second order.

Cows of the third order give thirteen quarts, and milk six months.

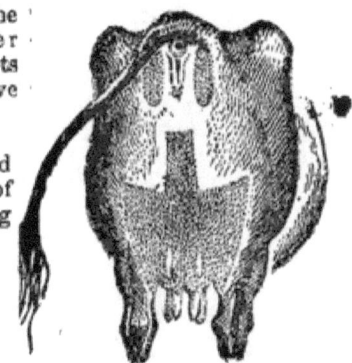

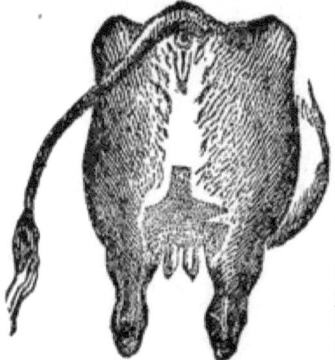

The fourth order has the escutcheon still smaller. The tufts alongside the vulva are not so plain, but the hair is coarse and bristly. There is a triangular invasion on the right of the escutcheon.

Cows of the fourth order give ten quarts and milk five months.

The Bastards of this class have a good escutcheon, but the tufts are large and of coarse, bristly hair, and will fail according to the size.

The Eighth Class. Square Escutcheon.

The first order of this class have the same yellowish color on the escutcheon as other first orders; the udder is flexible, covered with a short, fine down. The escutcheon is much of the shape of the Demijohns, but the vertical, as it rises, branches square off to the left, and ascends, straight and narrow, to the left side of the vulva. There are two ovals on the udder. The more the square approaches the vulva, and the finer the hair, the greater quantity is there of milk. Cows of the first order give seventeen quarts, and milk eight months.

The second order have a similar escutcheon, only smaller in every way. They have two ovals on the udder, and a small tuft to the right of the vulva.

Cows of the second order give fifteen quarts, and milk seven months.

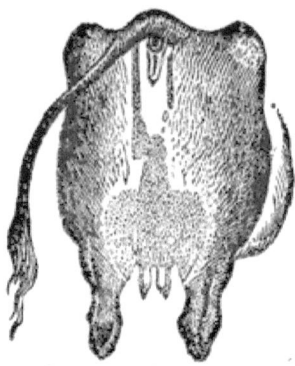

The third order have the escutcheon still smaller; the lines curving downward at the corners; one oval on the udder, and the tuft to the right of the vulva, larger and of coarser hair.

Cows of the third order give thirteen quarts, and milk six months.

The fourth order have the escutcheon much smaller, the square is much lower, and the upper part of it is formed of bristly hair, and feathery looking; as is also the tuft to the right. On the right side of the escutcheon is an invasion of triangular shape.

Cows of the fourth order give ten quarts, and milk five months.

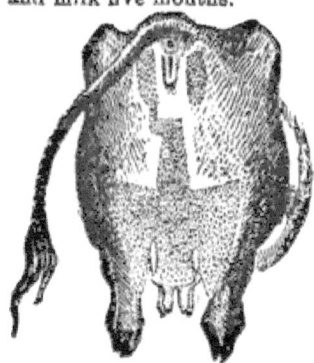

The Bastards of the eighth class are distinguished from those of the first order by the tuft on the right, which is of coarse and bristly hair, and the square terminates also in the same hair.

The Ninth Class. Limousines.

The escutcheons of this class, in ascending toward the vulva, do so in the shape of a spire, but stop short about half way.

The first order has the escutcheon of yellowish color, with flexible udder, covered with hair downy and silky. The shape is the same as the seventh and eighth class, except that the vertical escutcheon ends in a sharp point, like a spire or arrow head. There are tufts along each side of the vulva, and two ovals on the udder.

Cows of the first order give fifteen quarts, and milk eight months.

The second order is similar in the escutcheon, but smaller, with but one oval on the udder, and the vulva tufts larger, the left being longer than the right.

Cows of the second order give thirteen quarts, and milk seven months.

The third order is again smaller; the corners rounded downward; the tufts larger; no oval on the udder.

Cows of the third order give ten quarts, and milk six months.

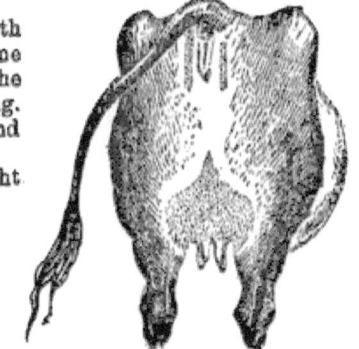

The fourth order same shape as the last, but still smaller and more rounding. The vulva tufts are of bristly hair, and the left one is seven inches long.

Cows of the fourth order give eight quarts, and milk five months.

The Bastards of the ninth class have a good escutcheon, but are distinguished by the large tufts of coarse, bristly hair alongside the vulva.

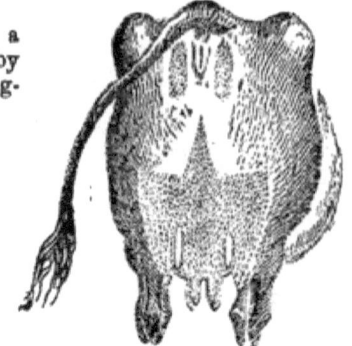

The Tenth Class.—Horizontal.

The first order have a dandruff of yellowish color; the hair is short, fine, and silky; the escutcheon is lower down from the vulva than the other classes, and is cut off by a horizontal line. There are two ovals on the udder; and two tufts, about three and one half inches long, on the sides of the vulva.

Cows of the first order give thirteen quarts, and milk eight months.

The second order has a smaller escutcheon; the vulva tufts are larger, the right shorter than the left; there is but one udder oval; in several of the orders of this class there is a small streak of ascending hair directly below the vulva.

Cows of the second order give ten quarts, and milk seven months.

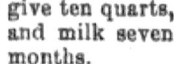

The third order have still smaller escutcheons; larger vulva tufts, the one on the left of bristling hair, four to five inches in length. The descending hair encroaches on the escutcheon on the right in a triangular shape.

Cows of the third order give eight quarts, and milk six months.

EFFECT OF CROSSING TWO ESCUTCHEONS.

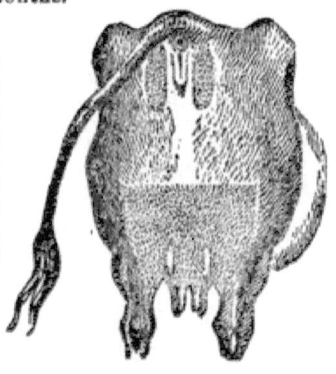

The fourth order have still smaller escutcheons; larger and coarser vulva tufts; and an invasion on the escutcheon on the right in triangular shape, and on the left of semi-circular shape.

Cows of the fourth order give seven quarts, and milk five months.

The bastards of the tenth class have the escutcheons large and good shaped; but are distinguished by the tufts alongside the vulva, these tell how long she will milk, by their size and the grade of the hair on them.

Effects of Crossing two Escutcheons.

CROSS BETWEEN THE SELVAGE AND LEFT FLANDERS.—The cows bearing this character are easily recognizable in certain breeds, and notably on those of the north-east of France.

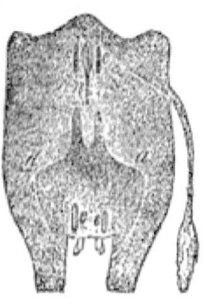

CROSS BETWEEN THE BICORN AND SELVAGE.—The *Epi*, or tuft, which I termed *jonctif*, or mesian tuft, and which is seen adhering under the vulva, is a favorable sign, and can be met with in those classes where the escutcheon does not reach as high as the vulva.

Cows which bear one or the other of these two escutcheons, are generally good milkers, and preserve their milk as well as cows of the first order of each class. These are the new escutcheons alluded to in Guenon's introduction, showing the effect of crossing.

ESCUTCHEONS ON BULLS.

Guenon applied his discovery to bulls to great advantage. He found that bulls belonged to the same classes as cows, and had escutcheons similar, but much smaller; these extend from the testicles upward toward the anus. The importance of having a good bull becomes apparent when we reflect that he "gets" from fifty to one hundred, annually, while the cow is impregnated but once in the year. The escutcheons of the progeny of a cow with good escutcheon will be much improved if the cow is coupled with a bull well marked, and particularly if his escutcheon is the same as that of the cow. Better have the two of different breeds, but of similar or good escutcheons, than to have the bull with inferior escutcheon.

Of course, the higher up the escutcheon extends on a bull, and the broader it is, the better it is, but we must not look for bulls to be so well-marked as cows are, for they never are. To distinguish the bastard bulls from the good ones, observe if there are any streaks of descending hair, and mixing so as to be bristly. This indication will be a certain one in proportion to the size of the blemish, and as that is in proportion to the whole escutcheon.

Guenon says: After having described, as I have done, all the classes of cows, and taught to recognize the bastards, I pass to the characteristic signs of the bulls reproducers, which can also be divided into orders and classes; the signs are the same as for the females, but they are much more restricted and of less extent.

With the males the escutcheon commences on the inside below the hams and extends as far as the middle of the posterior surface of the leg, and extends sometimes even to the anus of the superior orders in certain classes.

Like that of cows, the escutcheon of bulls is modified by tufts.

The bulls whose escutcheons are similar in their form and size to cows of the first order, possess a great ability for procreating good milk cows, those on the contrary whose escutcheons are but little developed, produce only those of poor yield.

A bull will be well marked, and a good reproducer when there is no interruption of descending hair in the ascending hair on the escutcheon; when the shape of the escutcheon is of large dimensions in proportion to the size of the animal, and it is covered with very fine hair.

The bulls of which the escutcheon is small and covered with coarse hair and irregular on the sides procreate bad milk cows, which give serous milk.

All interruptions in the ascending hair of the escutcheon by encroach-

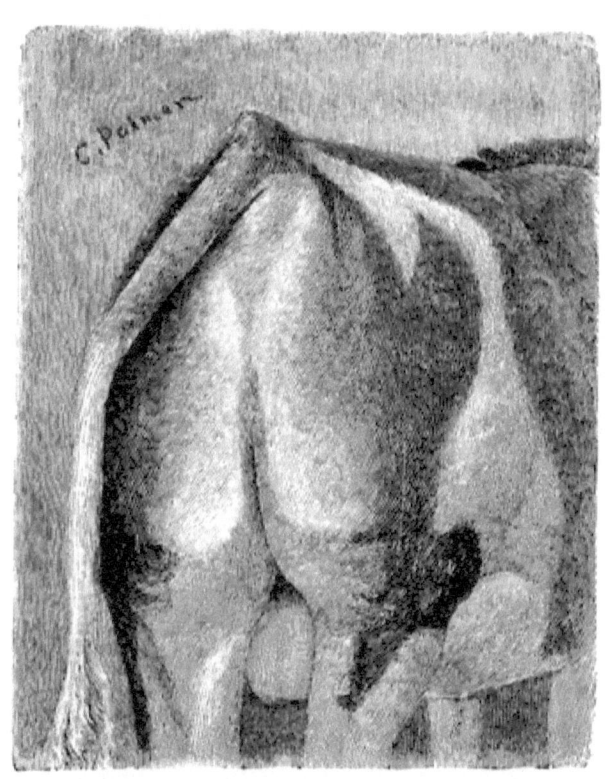

Escutcheon of Guernsey Bull RADLEY,
No. 209, A. G. H. B.

ments of the descending hair on the right or left, in the lower part of the thigh, indicate for their get a lower grade, and at a glance the inferiority of the milk production.

The yellow or nankeen color of the skin of the escutcheon is always a favorable sign.

The good reproducing bull will prove fecund until ten or fifteen years of age, but it is a rare exception.

Any one may be grossly deceived if he judges only by the appearance or the shape of the prolific qualities of a bull. Experience or observation alone can show that he has maintained his early ability.

A vigorous bull, well fed, can serve one or more cows each day, but it is of great importance that he shall not commence to serve until he is fifteen or eighteen months old, otherwise he will be speedily exhausted and deformed. The improving mark of his cross and his vigor will be speedily shortened.

When the bull has attained the age of two and a half or three years, the form alters, the hind-quarters become attenuated, the front quarter becomes much enlarged, his neck enlarges and thickens, &c.

About this time, whether he is castrated or whether he is "twisted," he preserves always the altered form of the bull, and is less sought for work, and in less request for butchering.

When the operations of castration and twisting are done too late, the animal has less predisposition to fatten; his flesh is harder and tougher; he is, however, in appearance in the same conditions of age, of quality, and of nourishment, as those castrated earlier.

Often bulls, whose character is docile and gentle, become wild and furious when they are used to serve.

In certain regions, to tame them, they put a ring of iron in their nose; in others, where the good use of these rings is not known, they are obliged to castrate or twist them. This operation suffices, generally, to control their passion; but, if not, they are sent to the butcher.

Classification of Bull Reproducers.

There are for bulls, as for cows, ten classes or families; each class subdivided into several orders, and each order comprises three grades, high, medium, and low.

I only admit, in each class, three orders. If one wishes to proceed in the application with more rigor, he will follow the sub-divisions of the classification of the cows. I will designate the three orders of each class by the denominations of good, medium, and bad.

The signs indicating the qualities which render the bull likely to beget good milk cows are placed, like those of the female, on the posterior parts. They start from the bag, and rise up to the anus, covering, also, the genital parts, and the scrotum.

With bulls, the escutcheons start from the anterior part of the bag, extending inside, and upon the hams, projecting on the thighs; from there, the curved lines, obtuse or acute, following the class, joining to the right or the left under the anus.

The escutcheon, in all its extent, is shown by the fineness of the hair, and the skin; by the color, more or less yellow, of the epidermis, and of the particles of dandruff which can be detached.

The characteristic secondary signs of the females will also be found in the males.

Bulls, like cows, have four and, sometimes, six false teats, which are found before the bag, in the direction of the navel. These teats are small and short.

Starting from the bag, one notices to the right and the left of the stomach two veins resembling the two milk veins of cows. They are prolonged to and pass a little in the direction of the navel, and terminate in a small cavity.

Independent of the characteristic signs indicated above, the bull reproducers should unite all the essential conditions which in each locality constitute the type of the pure race. These conditions are:

1. The color of the hide preferred in that country.
2. A size proportioned to the race that they are to continue. A shape and a frame usually accepted.
3. To be of the first order in each class, easily showing the power of transmitting milking qualities.
4. Aptitude for fattening.
5. To be good for work.
6. To have a docile and patient character.

The evils of conformation, like the good qualities, are transmitted generally by the act of generation. If it does not have the ability to do this, one should quickly correct it.

Here the bovine race has been much neglected in all these respects. A judicious choice, and a scrupulous attention is not always prevalent in selecting a breeding animal; thus it results in a fatal re-generation, to which it is time to put a stop.

Before giving the distinctive characteristics of the ten classes of bulls, it will be useful to mention those classes which are oftener met in French and foreign races; and also those which are more rare.

The classes which are most distributed, and which present the greatest number of bulls, are in all races these three classes: 1. The Curved-line; 2. The Limousine; 3. The Horizontal.

The classes on the contrary, which present but a very small number of subjects, are in the following order:

4. Demijohn.
5. Bicorne.
6. Square-cut.
7. Selvage.
8. Left Flanders.
9. Double Selvage.
10. Flanders.

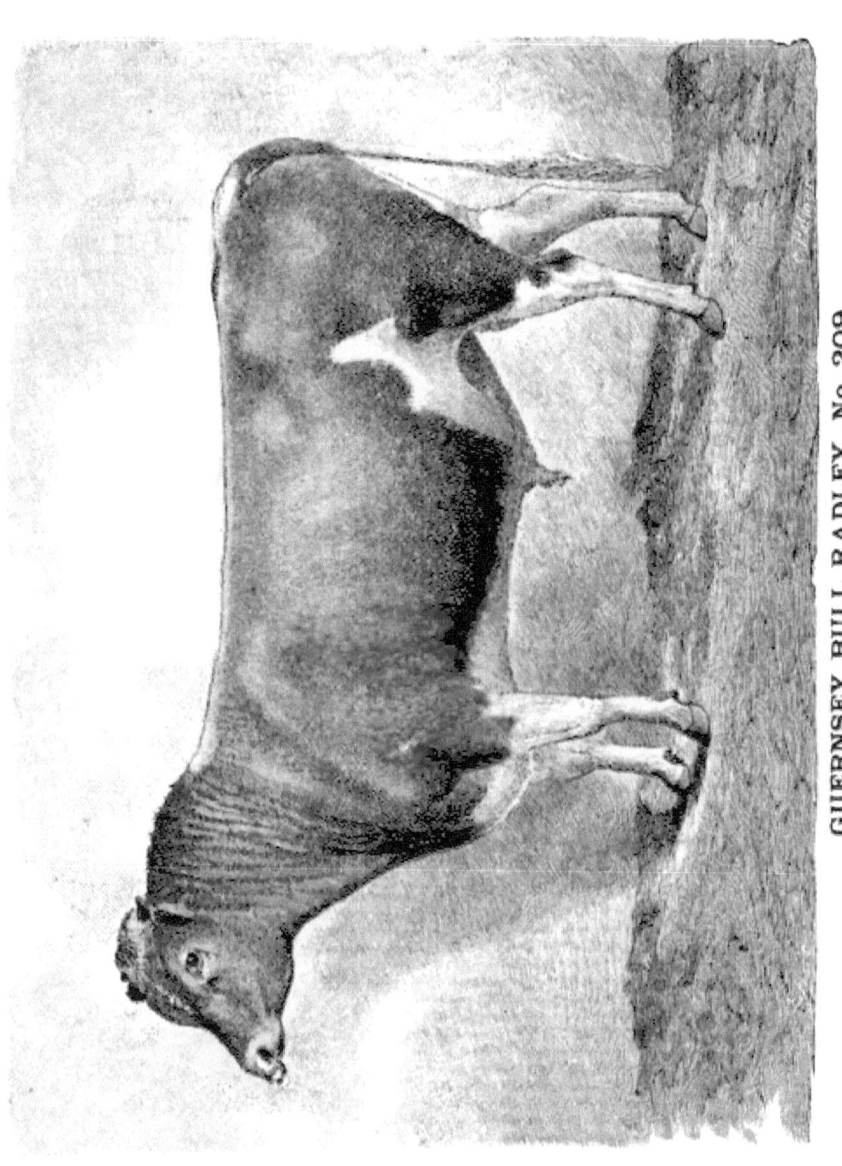

GUERNSEY BULL RADLEY, No. 209.
Property of S. C. Kent, West Grove, Pa.

The reason one finds so few good breeding bulls belonging to the first class, is first, the small number of such animals compared to that of cows; and next, the lack of knowledge of the best ones to keep. Oftentimes for want of this knowledge, the best bulls were castrated for oxen or for fattening, thus by chance, the poorest are often kept.

The best individuals have generally at birth, all the qualities which characterize a superior animal. They are easily kept and fattened, for the reason that their mother has much milk, and are soon ready for the butcher. Inferior animals, on account of a smaller supply of milk, are thin, and often malicious, of little value, and remain oftener in the hands of the owner. Thus are sacrificed the good bulls, and the bad are kept. Therefore, always select the choicest when they are young, to improve the race.

It will thus be seen, Guenon divided his bulls into three classes: The good, the mediocre, and the bad. He also divided them into three sizes: The high, the medium, and the low. But he makes no difference between the three sizes of bulls in his description of the escutcheon. He describes each one of the three principal orders, leaving to the practitioner to determine the intermediate degrees between the good and the mediocre, and between the mediocre and the bad.

We do not repeat his descriptions, as they are based upon those of the cows of the same classes, and the engravings tell the whole story. We reproduce the engravings of the good and mediocre. But very occasionally is one of the " rare" ones observed, but he says the Curved-line is the most usual, then the Limousine, and lastly the Horizontal. What we give is quite sufficient for all practical purposes. We advise all to carefully select their breeding animal, which will, in most cases, be from among what he calls the " mediocre."

Class I.—Flanders Bull.

Class II.—Left Flanders Bull.

Class III.—Selvage Bulls.

CLASSIFICATION OF BULLS.

Class IV.—Curveline Bulls.

Class V.—Bicorn Bulls.

 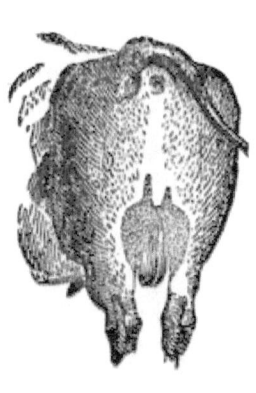

Class VI.—Double Selvage Bulls.

 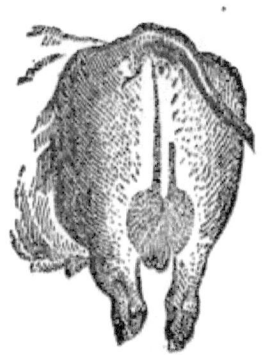

Class VII.—Demijohn Bulls.

Class VIII.—Square Bulls.

Class IX.—Limousine Bulls.

REPORT OF THE PENNSYLVANIA GUENON COMMISSION.

At the annual meeting of the Board held January 2, 1878, it was
"*Resolved*, That the president of the Board (His Excellency Governor John F. Hartranft) be authorized and requested to appoint a commission of experts, who shall inquire into and report upon the reliability of the Guenon or escutcheon theory for selecting milking stock; said report to be made to the secretary of the Board on or before the 1st of November next."

In accordance with this request, His Excellency Governor Hartranft issued the following commission:

COMMONWEALTH OF PENNSYLVANIA,
EXECUTIVE CHAMBER, HARRISBURG, *April 24, 1878.*

To GEORGE BLIGHT, Esq., *of the city of Philadelphia;* CHALKLEY HARVEY, Esq., *of the county of Delaware, and* WILLIS P. HAZARD, Esq., *of the county of Chester:*

GENTLEMEN: I have the honor to inform you that you have been duly appointed a committee by the State Board of Agriculture to investigate and test "The Guenon Milk Escutcheon Theory," and report the result thereof to the secretary of said Board.

JNO. F. HARTRANFT,
Governor and President of the Board.

November 1, 1878.

To the Honorable JNO. F. HARTRANFT,
Governor and President of the Board of Agriculture:

YOUR EXCELLENCY: In compliance with the commission tendered us, we beg leave most respectfully to report that we have visited a number of herds and have examined two hundred cows, the result of which is herewith submitted.

Having performed to the best of our ability the duty assigned us, we beg leave to be discharged from further consideration of the question.

Respectfully yours,

GEORGE BLIGHT, *Philadelphia,*
CHALKLEY HARVEY, *Chad's Ford,*
WILLIS P. HAZARD, *West Chester.*

The Pennsylvania Guenon Commission having been appointed "to investigate and test the Guenon or Milk Escutcheon theory, and report the result thereof," respectfully report that they have examined two hundred cows, heifers, and bulls, and the result of their examinations has been to convince themselves and others of the merits of the system, of its exceeding value to the practical farmer; and they believe that if generally followed for twenty years, the value of the neat cattle of the State would be increased vastly, the amount of milk and butter produced would be much larger, and the quality of both articles better, while the quality of the meat would be improved. Having believed in and practised the system for many years, they would add that their recent extended and careful examinations and contact with a number of owners of all grades of stock, has tended to confirm them more thoroughly in their belief. As an adjunct to previous knowledge to assist purchasers or breeders of cattle in getting or raising the best, and weeding out the poorest, they think it is worthy of being acquired by every farmer. And they would recommend their fellow farmers not to be dismayed at the apparent difficulties to be surmounted in obtaining a knowledge of the system, as it is only absolutely necessary to acquire a knowledge of the first four orders of each class, and a few other points, to practically apply it, as all animals below those grades are not worthy of being kept. Any intelligent man can readily master the system, and soon become proficient in it by practice. This knowledge, applied with the tests heretofore usually used, will enable any one to become a good judge of cattle.

The manner of making up their account of each animal is to examine the escutcheon and the udder, from which they place her in the class and order nearest to those delineated by Guénon, and then estimate the quantity, quality, and time that she will milk. These estimates must be, of course, only *approximate*, as they are based upon the indications of the escutcheon, the size of the cow, and her probable condition. As it is readily seen that where estimates are based upon what the cow should do *within three months of her being fresh*, it would be impossible to always grade the exact value of all the cows in a herd, each of which is at a different period of gestation, or in a different condition or state of health, and where also the cow is affected by the way in which she is fed and cared for, by the season, by the state of the temperature, and other circumstances. The estimates are based upon what the commission thinks the cow would do when all the conditions are favorable to her development, and where she is properly fed and cared for. A record is made by the commission on the spot. An account of the qualities of each head is drawn up by the owner. Each is made at separate times, and without the knowledge of the other party. Then the two accounts are copied off into parallel columns for comparison. If the accounts agree in seventy-five per cent. out of one hundred, it certainly must be presumed the system has sufficient value to make it worthy of adoption by all farmers and breeders. As every farmer knows the yield is much influenced by the feed, the care, the exposure, and the treatment of the cows; therefore, a certain amount of allowance must be made, for these various things will so alter matters, that no one can tell to a quart, or a pound of butter, or to the week in time of milking. In fact, every farmer knows neither the owner himself, nor his man, can tell to a quart how much his cow or cows actually give, unless a daily record is kept every day of every year. For even if he does keep such a record, he will find the various circumstances named above affecting the quantities in his record. Therefore the earnest seeker after truth, comparing the statements made in the two columns, must not expect the two to tally without some variations. The true spirit with which he must examine these statements, will suggest itself in the question: Is this a system by which I can judge of the value and quantities of a cow correctly? Is this a system that will tell me the points of a cow, good or bad, more correctly than by any other method? Let the candid inquirer weigh these statements, and think if he knows of any method by which he can go into a herd and surely pick out the best cows, and leave the poor ones to those who judge not by this system. Every farmer has his own mode of judging, but take the shrewdest and most practiced, can he avoid often the bastards? What the commission find they can do, is that in a large

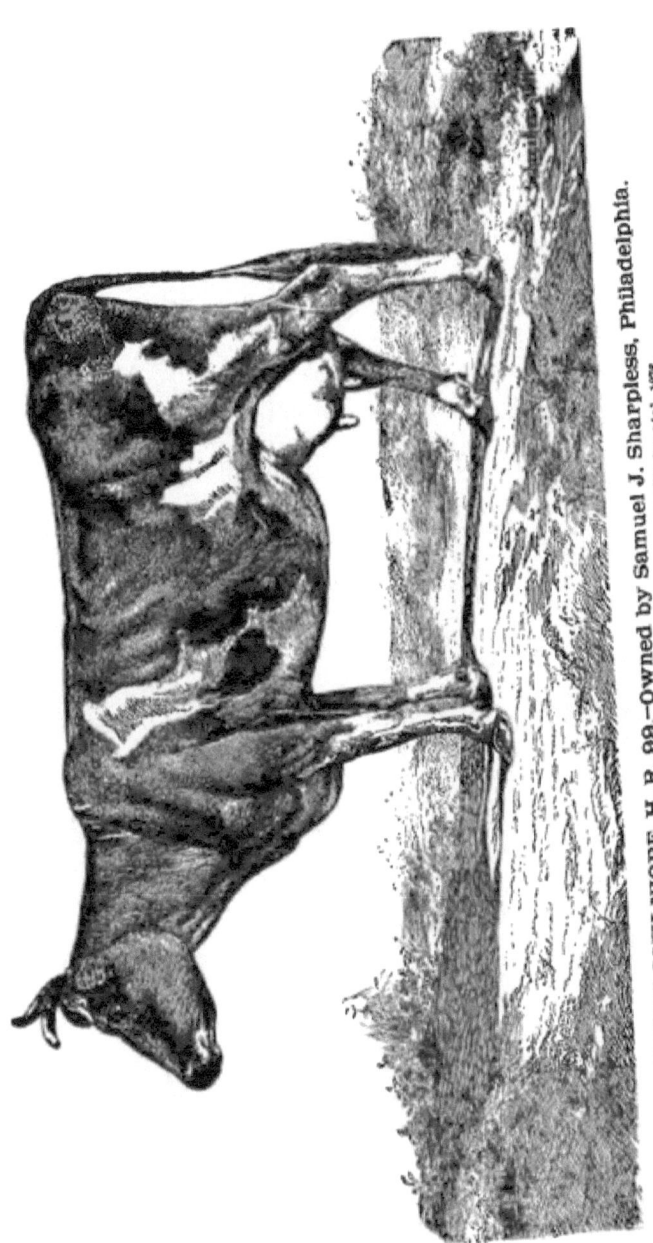

JERSEY COW NIOBE, H. R. 98.—Owned by Samuel J. Sharpless, Philadelphia.
Was awarded First Prize as the BEST Cow at the Centennial, 1876.

majority of the percentage of cases, they will give a good estimate of the qualities of any animal. Their opinions of the time a cow will go, is based upon what they think should be the treatment of all cows, viz.: that every cow should have a rest of from four to six weeks, at least.

The Commission at Barney's Farm.

The members of the Guenon commission, visited the farm of John B. Barney, on the 9th of May, 1878, and examined twelve cows, mostly Grade Durhams, Grade Jerseys, and farm stock, and they were uniformly successful in judging of said stock, with some difference of opinion on two of them.

"I was present at the examination of twelve cows of my herd, and think the committee were so uniformly successful in judging of the merits of the different cows, with such slight variations of opinion between us, as to increase my belief in the Guénon system being of great advantage to the farmers in selecting stock.

JOHN B. BARNEY.
Chadd's Ford, Chester county."

May 16, 1878.

The Commission at Sharpless' Farm.

The commission visited the fine farm and herd of Jersey cows of Samuel J. Sharpless, at Street Road station, West Chester railroad, May 10. Present. Messrs. Harvey, Blight, Hazard, and Thomas J. Edge.

SAM'L J. SHARPLESS' HERD OF JERSEYS, AS REPORTED BY E. J. DURNALL, HERDSMAN FOR S. J. S., May 10, 1878.	SAM'L J. SHARPLESS' HERD AS REPORTED UPON BY GUENON COMMISSION OF THE STATE, MAY 10, 1878.
No. 1.—Seven years. Quantity, about 12 quarts a day. Quality, medium. Milks about 10 months.	*No. 1.*—Curveline cow, second order. Quantity, if 14 quarts, doing well. Quality, good. Will milk ten months out of twelve.
No. 2.—Thirteen years. Quantity, best. Gives 24 quarts 3 months after calving. Quality, second rate. Has made 11¼ pounds in a week. Milks full up to time, except when she had twin calves.	*No. 2.*—Flanders cow, third order. Quantity, 16 quarts first three months. Quality, third rate. Dry two months.
No. 3.—Eleven years. Quantity, medium. Quality, best. Would go to her time.	*No. 3.*—Flanders cow, third order. Quantity, 12 quarts; three months. Quality, very good and rich. Dry six weeks.
No. 4.—Eight years. Quantity, medium. Quality, good; makes about 10 pounds. Up to her time.	*No. 4.*—Flanders cow, second order. Quantity, 14 quarts. Quality, very fine. Well up to her time.
No. 5.—Imported. Eleven years. Quantity, about 17 to 18 quarts a day. Quality, best; makes 11 pounds per week. Up to her time.	*No 5.*—Selvage cow, first order. Quantity, best; about 18 quarts. Quality, no question. Milks up to her time.
No. 6.—Ten years. Quantity, second rate. Quality, medium. About two months short of her time.	*No. 6.*—Flanders cow, first order. Superior milker. Quality, second class. Milks up to her time; say six weeks.
No. 7.—Two years old. Had only first calf. Quantity, medium. Quality, good. Not fairly tested for time.	*No. 7.*—Curveline cow, second order. Quantity, medium. Quality, too young for quality; say good. Time, too young for test.
No. 8.—Four years. Quantity, medium. Quality, first class. Up to calving.	*No. 8.*—Selvage cow, second order. Quantity, medium. Quality, good. Up to her time; say six weeks.
No. 9.—From Niobe Third. Three years. Quantity, first rate. Quality, first rate. Up to her time.	*No. 9.*—Flanders cow, second order. Quantity, first class. Quality, first class. Well up to her time.

No. 10.—Imported. Four years. Had first calf at Centennial, in October, and made in seven days, 9 pounds 10 ounces.
Quantity, about 16 quarts.
Quality, excellent.
Up to time. Has been milking two years.
No. 11.—Ten years.
Quantity, second highest of herd; best.
Quality, second class. Makes about 10 pounds.
Full up to her time.
No. 12.—Four years.
Quantity, second rate.
Quality, second rate; about 7 pounds.
Milks to three months of her time.
No. 13.—Six years.
Quantity, number one.
Quality, number one.
Full up to time.
At seven months from calf gives 16 quarts.
No. 14.—Five years.
Quantity, promises fair.
Quality, good.
No. 15.—Four years. Of Niobe stock, the poorest.
Quantity, third rate; 6 quarts.
Quality, good; second rate.
Up to her time.
Dropped her calf.
No. 16.—Quantity, number one.
Quality, number one.
Up to her time.
No. 17.—First calf.
Quantity, number one.
Quality, number one.

No. 10.—Decided to pass her.

No. 11.—Horizontal cow.
Quantity, first-class.
Quality, inferior.
Milk up to eight months.
No. 12.—Flanders cow, third order.
Quantity, light.
Quality, third class.
Three months short of her time.
No. 13.—Flanders cow, number two order.
Quantity, second class.
Quality, first class.
Up to her time.

No. 14.—Flanders cow, first order.
Quantity, good.
Quality, fair.
Within a month of her time.
No. 15.—Flanders cow, second order.
Quantity, about 12 quarts.
Quality, not very fine.

No. 16.—Selvage cow, first order.
First class every way.

No. 17.—Flanders cow, second order.
Quantity and quality, fair.

The commission and Mr. Durnall agree as to the best cow, selected from the first six on this list—on the one side by the marks, and on the other from his knowledge.

"Having compared the annexed account of the qualities of the seventeen cows of my herd, examined by the State Guenon Commission, with the originals of the accounts as given by both parties at separate times, and taken down upon the spot, I believe it to be a true and faithful transcript of the original record of the examination.
SAMUEL J. SHARPLESS."
PHILADELPHIA, *May 20, 1878.*

"I was present at the examination of Mr. Sharpless' herd of Jerseys, made on the 10th of May by the State Guenon Commission, and having examined the accounts of the herd given by me, as hereto annexed, with the original entries of those given by me, and also the accounts of the commission, with the original written opinions, do certify that the annexed are faithful transcripts of the records made by each party at separate times, and that the statements were unknown to each other.
E. J. DURNALL,
Herdsman for Samuel J. Sharpless."
LENAPE FARM, *May 20, 1878.*

The Commission at Strode's Farm.

The members of the Guenon Commission visited the dairy farm of Marshall Strode & Son, who have a large butter factory, and are celebrated for their first-class butter, and they examined seventeen head of grade dairy stock, and according to the testimony of Mr. Strode, who accompanied them in their examination, were successful in judging according to the Guenon system, fifteen cows out of seventeen examined. Viewed May 10.

Present, Messrs. Harvey, Blight, Hazard, and Edge.

REPORT OF PENNSYLVANIA COMMISSION.

"Having been present when the members of the Guenon Commission examined seventeen of our herd, and having witnessed the accuracy with which they determined the quality of the stock inspected, we bear testimony to the fact that their judgment was correct, according to our experience with the cows, in fifteen cases out of seventeen, and even in these two they were partially successful. And we are more confirmed in our previous belief in the value of the system, as we never buy a cow for a good one that is not well marked. We run a dairy of seventy-one cows.

Yours truly,

MARSHALL STRODE & SONS."

EAST BRADFORD, *May 15, 1878.*

Examination of Thomas M. Harvey's Stock of Jerseys and Guernseys, May 11, 1878.*

This herd is one of the finest in the State. The cows are kept in good condition, and being well fed, the yield is very large per head. Their product is first class butter, and should bring the highest price in the market.

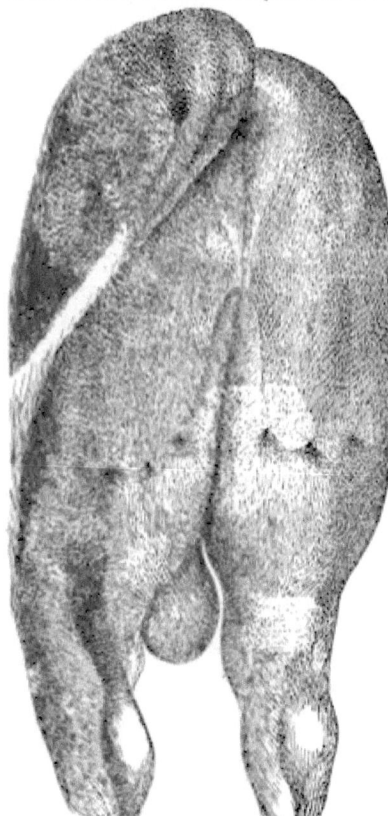

Escutcheon of Imported Guernsey Bull "Sir Champion."

The commission examined, in addition to the twenty-five cows on this list, Mr. Harvey's Guernsey imported bull "Sir Champion," which is thoroughly well marked; perhaps, the best marked bull in the country. The value of his get is, therefore, very decided. It shows most conclusively the importance of a bull from good milking stock, and that he should have a good escutcheon. The importance of a good sire to stamp his qualities upon his descendants was conclusively proved by Mr. Harvey's younger stock. The commission examined a young bull of seven months age, which was as perfectly and beautifully marked as his sire, and as nearly the same shape escutcheon as his sire's as could possibly be. Mr. Harvey has since sold him for $100, to Colonel R. M. Hoe.

Nos. 9½ and 9¾ prove also the gain to a herd from a well marked bull. These were yearlings of Champion's get. No. 9½ was a great improvement upon the mother, No. 9, Carrie, of this list.

In the statements of the commission as to quantity, they have not mentioned the number of quarts, as the amounts given by most of this herd are superior to the generality, even of Guernseys; and, owing to good selection and careful handling by their owner, the commission judge that the number of quarts would be larger than usual.

---

* In this examination two cows which had already been examined and reported upon by the commission were, without their knowledge, afterwards brought up for another examination, in which their opinion as recorded, agreed almost exactly with the one previously recorded, thus affording a strong proof of the value of the system. See reports of No. 1, Betsy, and Nos. 4 and 20, Beauty. (Secretary of Board of Agriculture.)

STOCK OF THOMAS M. HARVEY. THOMAS M. HARVEY'S STATEMENT.

No. 1.—BETSEY.

Quality, medium.
Quantity, 17 quarts.
Never dry.

No. 2.—NANCY.

Quality, first rate.
Quantity, 16 quarts.
Will milk up to calving.

No. 3.—CLAUDE.

Quality, first rate.
Quantity, at first, 18 quarts; but at six months, 8 quarts.
Not inclined to go dry altogether.

No. 4.—BEAUTY.—(Worth.)

Quality, first rate.
Quantity, 20 quarts.
Almost impossible to dry off.

No. 5.—ZILLA.

Quality, medium.*
Quantity, 18 quarts.
Never dry.

No. 6.—CHERRY.
Quality, good.

Quantity, 24 quarts.
Will milk on.

No. 7.—ECHO.

Quality, medium.
Quantity, 20 quarts.
Does not go dry.

No. 8.—MINNA.—Guernsey.

Quality, first rate.
Quantity, 17 quarts.
Not inclined to dry.

No. 9.—CARRIE.

Quality, first rate.
Qantity, 10 quarts.
Will go dry two months.

No. 10.—GENTLE.—Jersey.
Quality, first rate.
Quantity, 20 quarts.
Dry two months.

No. 11.—MARY.—Grade, Durham.

Quality, medium.
Quantity, 18 quarts.
Dry two months.

No. 12.—EUGENIE.—Jersey.
Quality, medium.*
Quantity, 18 quarts.
Does not dry.

No. 13.—VICTORIA.—Guernsey.
Quality, medium.*
Quantity, 20 quarts.
No drying.

No. 14.—JENNIE.
Quality, good.
Quantity, 20 quarts.
One month dry.

STOCK OF THOMAS M. HARVEY. OPINION OF THE GUENON COMMISSION.

No. 1.—BETSEY.—Jersey. Curveline, No. 1.
Quality, first rate.
Quantity, first rate.
Time of milking; up to her time.
Second examination confirmed.

No. 2.—NANCY.—Guernsey. Flanders No. 2.
Quality, first rate.
Quantity, first rate.
Six weeks.

No. 3.—CLAUDE.—Guernsey. Curveline, 3d.
Quality, first-class for rich milk.
Quantity, not large.
Dry three months.

No. 4.—BEAUTY.—(Worth.) Curveline, 1st.
Quality, first class.
Quantity, large.
Close to time.
Reexamined—See report No. 20.

No. 5.—ZILLA.—Guernsey grade.
Imperfect Selvage of low order. Irregularly marked.
Will milk well for few months only.
A good shaped cow.

No. 6.—CHERRY.—Half Jersey.
Good milk.
Milks profusely, and keeps herself thin on account of it.

No. 7.—ECHO.—Grade, Jersey. Selvage, 2d.
Quality, medium.
Quantity, medium.
Milks up pretty well.

No. 8.—MINNA.—Guernsey. Curveline, 2d.
Quality, good.
Quantity, good milker until within two months of calving.

No. 9.—CARRIE.—Strangely and imperfectly marked.
Quality, good.
Quantity, poor.
Dry up soon.

No. 10.—GENTLE.—Jersey. Selvage, 1.
Quality, first rate.
Quantity, large.
Dry six weeks.

No. 11.—MARY.—Grade, Durham. Flanders, 1.
Quality, first class.
Quantity, good.
Dry six weeks.

No. 12.—EUGENIE.—Jersey. Curveline, 2d.
Quality, first rate.
Quantity, large.
Dry two months

No. 13.—VICTORIA.—Guernsey. Selvage.
Quality, medium.
Quantity, good while she milks.
Dry two months.

No. 14.—JENNIE. Ordinary stock.
Quality, not very rich, but good.
Quantity, good.
Dry two months.

No. 15.—MAGGIE, 2d—Young.

  Quality, medium.*
  Quantity, 10 quarts.
  Dry two months.
No. 16.—ROCKET.

  Quality, medium.
  Quantity, 15, 20, to 25 quarts; variable.
  Dry three months.
No. 17.—AMY.
  Quality, good.*
  Quantity, 14 quarts.
  Ten weeks dry.
No. 18.—COMLY.
  Quality, first rate.
  Quantity, 15 quarts.
  Dry two months.
No. 19.—KITTY.

  Quality, first rate.
  Quantity, 16 quarts.
  Dry six weeks.
No. 20.—BEAUTY.

Second examination.
See No. 4.
No. 21.—ROSETTA.—Guernsey, imported.

  Quality, first rate.*
  Quantity, 18 quarts.
  Does not go dry.
No. 22.—DUCHESS.—Guernsey.

  Quality, first rate.
  Quantity, 20 quarts.
  Don't go dry.
No. 23.—BRIDGET.—Guernsey grade.

  Quality, first rate.
  Quantity, 14 to 18 quarts.
  Fails when pregnant. Does not dry altogether.
No. 24.—FANCY.—Guernsey.
  Quality, medium.
  Quantity, 16 quarts, and fails fast.
  Dry three months.
No. 25.—DAFFY.—Grade, Durham.

  Quality, medium.
  Quantity, 22 quarts.
  Does not dry.

No. 15.—MAGGIE, 2d.—Guernsey. Flanders, 3d.
  Quality, very good.
  Quantity, medium and continuous.
  Dry two months.
No. 16.—ROCKET.—Grade, Jersey. Curveline, 2d.
  Quality, rich.
  Quantity, good.
  Dry two months.
No. 17.—AMY.—Kentucky Short-Horn.
  Quality, medium.
  Quantity, poor.
  Dry three months.
No. 18.—COMLY.—Guernsey. Selvage, 3d.
  Quality, first rate.
  Quantity, good and continuous.
  Dry six weeks.
No. 19.—KITTY.—Guernsey and Jersey grades. Bicorn, 2d.
  Quality, first class.
  Quantity, first rate.
  Milks well up.
No. 20.—BEAUTY.—Guernsey. Curveline, 1st.
  Quality, first class.
  Quantity, first class. Milks well up.
  This is the second examination.
  See No. 4.
No. 21.—ROSETTA.—Guernsey, imported. Flanders.
  Quality, excellent.
  Quantity, fair milker.
  Dry two months.
No. 22.—DUCHESS.—Guernsey, imported. Curveline, 2d.
  Quality, medium.
  Quantity, large.
  Dry two months.
No. 23.—BRIDGET. Guernsey grade. Selvage, 4th.
  Quality, good.
  Quantity, poor.
  Dry three months or ten weeks.

No. 24.—FANCY.—Guernsey. Selvage.
  Quality, medium.
  Quantity, poor.
  Dry two months.
No. 25.—DAFFY.—Grade, Durham. Flanders, 1st.
  Quality, good.
  Quantity, about 24 quarts.
  Milks her full time.

In the above account will be noticed a few with the mark * which Mr. Harvey said indicates that the amount of milk given and the quality was largely increased by extra feeding.

"I was present at the examination of our herd of Guernseys, Jerseys, and grades, by the Guenon Commission, on the eleventh of fifth month, 1878, and I have examined their report and compared it with the originals written on the spot, (the contents of which were before now unknown to me,) and I testify to the annexed report being an accurate copy of the opinions recorded by them at the time of examination. The statements made by me were handed to members of the commission on twenty-first instant, and are as nearly accurate as my own knowledge, and that of the persons who had the immediate care of the herd, and an interest in the proceeds of the dairy, could make them.

THOS. M. HARVEY."

WEST GROVE, *27th of 5th month, 1878.*

"Having had the immediate care of the herd of Thomas M. Harvey, and an interest in the proceeds of the dairy for six years, and being well acquainted with the merits of each member of the herd, I can certify that the statements we have made to the

Guenon Commission are correct to the best of our knowledge and belief, and were made more than a week after the commission had recorded their opinions; which opinions were unknown to us until the present time. I have also compared the record of opinions herewith with the originals, and find them to be a correct transcript of them.

CLARKSON MOORE."

"Having had the immediate care of the herd within alluded to, and an interest in the proceeds thereof for the past eleven months, I can certify that I united with Clarkson Moore in making the statement relative to the quality and productiveness of the cows examined by the Guenon commission. I was from home when they were here, and knew nothing of their opinions when we made up our statement.

ZEBEDEE HAINES."

Fifth month, 28th, 1878.

"I was present on Saturday, March 11, 1878, at the farm of Thomas M. Harvey, when the examination of that portion of his herd was made by the Guenon Commission, from Nos. 10 to 25, inclusive, of their report. I have compared the original recorded opinions of the commission with the report herewith, and find the latter correct.

N. J. SHARPLES,
President of the Experimental Farm Club."

Examination of William M. Large's Herd, Chestnut Grove, Doylestown, Bucks County.

The commission, as represented by George Blight and Willis P. Hazard, visited the beautiful farm of William M. Large, on the afternoon of May 31—on a very rainy, unpleasant day, and making the examination of stock a very difficult duty. The stock is a valuable one of thorough-bred and grade Short-Horns and is well fed and otherwise well cared for.

WM. M. LARGE'S ACCOUNT OF HIS HERD.	OPINIONS OF THE STATE GUENON COMMISSION.
No. 1.—VICTORIA.	*No. 1.—VICTORIA.*—Short-Horn. Flanders, first order.
Quantity, 18 quarts.	Quantity, about eighteen quarts.
Never tried her on butter but once, then made 10? pounds.	Quality, good.
Goes dry two months to ten weeks.	Up to her time—say one month.
No. 2.—JOSEPHINE.	*No. 2.—JOSEPHINE.*—Thorough-bred Short-Horn. Flanders, second order.
The Doylestown Agricultural Society offered a premium of $25 for the cow that yielded the most butter; and also $25 for the cow that gave the most milk. The largest yield for a single week 16 pounds 3 ounces. Was tried five times during the year, two months apart; taking the average of the five consecutive trials, would make a trifle over 500 pounds. Awarded the first premium for butter, and second for milk.	
	Quantity, eighteen quarts.
	Quality, very good.
Lost the record of pounds of milk.	About one month dry.
Dry about one month.	
No. 3.—FANNY FERN.	*No. 3.—FANNY FERN.*—Flanders, first order.
Quantity, nineteen quarts.	Quantity, twenty quarts.
Quality, never tried her butter production.	Quality, first rate.
Goes dry five or six weeks.	About up to time, (one month.)
No. 4.—LETTIE.	*No. 4.—LETTIE.*—Selvage, fifth order.
Quantity nine quarts.	Quantity, eight quarts.
Quality, never tried her butter qualities, but her milk is rich and good.	Quality, second rate.
Goes dry about three months.	Dry four months.
No. 5.—NORAH.	*No. 5.—NORAH.*—Flanders, imperfect.
Quantity, fresh, gives seventeen quarts; holds to it well.	Quantity, eighteen quarts.
Quality, has made 10 pounds butter in a week.	Quality, good.
Goes dry about two months.	Dry three months.

No. 6.—LUCY.
  Quantity, thirteen quarts.
  Quality, a less number of pounds will make a pound of butter than most of my other cows; think her milk extra good.
  Dry some two months.
No. 7.—BERNICE.
  Quantity, when fresh, about twelve quarts.
  Quality, never tried her butter production.
  Goes dry some three months, and has the credit of being the poorest cow in the herd.
No. 8.—JOSEPHINE 2d.—First calf.
  Quantity, first calf, eleven quarts.
  Quality, never weighed her milk or tried her butter production.
  Cannot tell how long she will go dry.
No. 9.—JUDITH.
  Quantity, seventeen quarts.
  Quality, on a trial after her first calf was taken away, made 2 ounces less than 8 pounds of butter, done up in lumps for market.
  Goes dry six or seven weeks.

No. 6.—LUCY.—Flanders, fourth.
  Quantity, fourteen quarts.
  Quality, good, very.

  Dry three months.
No. 7.—BERNICE.—Flanders, second.
  Quantity, twelve quarts.

  Quality, good.

  Time, eight months out of twelve.

No. 8.—JOSEPHINE 2d.—Curveline, third.
  Quantity, ten to twelve quarts.
  Quality, rich.

  Goes to two months of her time.
No. 9.—JUDITH.—Flanders, first.
  Quantity, eighteen quarts.
  Quality, good.

  Well up to her time (one month or six weeks.)

"Having been present at the examination of my herd of Short-Horns, by the State Guenon Commission, May 31, 1878, I certify this report is a correct copy of the original records made on the spot, and at separate times; neither party having knowledge of the other's accounts.

WILLIAM M. LARGE."

CHESTNUT GROVE, 7th month, 3d, 1878.

Report of the Examination of the Stock of Eastburn Reeder, of Rabbit Run Stock Farm, New Hope, Bucks County, Pa.

The commission visited the farm of Eastburn Reeder on Saturday, June 1, and examined his stock of Jersey and Guernsey breeds in presence of the proprietor and a number of members of the Solebury Farmers' Club. Mr. Reeder's account of his herd had been drawn up and placed in the hands of J. S. Williams, Esquire, secretary of the Solebury Farmers' Club, some ten days before the visit of the commission, and is printed herewith.

The commission was represented by George Blight and Willis P. Hazard, and part of the time by Captain J. C. Morris, of Susquehannna county, at the request of Thomas J. Edge, secretary of the State Board.

Mr. Reeder, not having made in his report any statements of the quality of the milk, except as regards the yield in butter, has sent the commission the following condensed statement of what each cow gave on May 20:

No. 1, Belle,	10 quarts.	No. 7, Florentia,	10 quarts.
No. 2, Topsy,	14 "	No. 8, Paunacussing,	8 "
No. 3, Firefly,	12 "	No. 9, Lady Delaware,	6 "
No. 4, Isabelle,	8 "		
No. 5, Marian,	14 "	Total for nine cows,	94 "
No. 6, Urania,	12 "		
Yield of the herd, May 20,	94 quarts.	Butter in seven days,	67 pounds.
Yield of the herd in seven days,	658 "	Quantity of milk to one pound of butter,	9¾⁴ qts.

This statement of the number of quarts required to make a pound of butter from Jersey and Guernsey stock, it will be seen, carries out the conclusions of the commission, who estimated the quality of this herd, as well fed and not too long milked stock of these breeds should give a pound of butter to every nine quarts of milk.

"Having been present at the examination of my herd by the State Guenon Commission, on Saturday, June 1, 1878, and having since examined their report by the original record made on the spot by them, and compared it with my account of the herd, handed to Mr. J. S. Williams, May 21, or more than one week before they made the examination, I do certify that the accompanying report is correct and true.
EASTBURN REEDER,
New Hope, Bucks county, Pa."

RABBIT RUN FARM, *June 15, 1878.*

"I certify that Eastburn Reeder handed me his account of his herd ten days before the examination was made by the State Guenon Commission; that I was present with others at the examination; that the two reports were compared in the presence of a number, shortly after the examination, and were generally satisfactory; and that I have now examined the accompanying reports by the two original records, made at separate times as above mentioned, and have found them correct and true copies of said original records.
J. S. WILLIAMS,
Secretary of the Solebury Farmers' Club."

June 15, 1878.

Examination of Eastburn Reeder's Herd.

ACCOUNTS OF THE HERD BY EASTBURN REEDER.	OPINIONS OF THE HERD BY THE GUENON COMMISSION.
No. 1.—BELL.—Age, 12 years. Grade, Alderney. Calved November 10, 1877. Greatest yield when fresh, 16 quarts per day. Yield May 20, 1878, 10 quarts per day. Butter, 8 pounds in seven days. Tried February, 1875. Milks to one month of calving.	No. 1.—BELL.—Grade, Alderney. Flanders, 2d. Quantity, 18 to 20 quarts. Quality, fair. Dry three to four months.
No. 2.—TOPSY.—Age, 10 years. Guernsey. Calved March 9, 1878. Greatest yield when fresh, 18 quarts. Yield May 20, 1878, 14 quarts. Made 12 pounds in seven days. Tried June, 1872. Goes dry three months before calving.	No. 2.—TOPSY.—Guernsey. Curveline, 2d. Quantity, 16 to 18 quarts. Quality, first rate. Dry two months.
No. 3.—FIREFLY, (1133.)—Age, 8 years. Jersey. Calved October 28, 1877. Greatest yield when fresh, 14 quarts. Yield May 29, 1878, 12 quarts. Averaged 6 pounds butter for forty weeks, from Sept. 1, 1872, to June 8, 1873. Greatest yield of butter in any one week since, 10¼ pounds. Never goes dry; has been milked regularly since August 27, 1872.	No. 3.—FIREFLY.—Jersey thorough-bred. Demijohn, 2d. Daughter of Niobe, 3d. Quantity, 12 to 14 quarts. Quality, medium. Dry two months.
No. 4.—ISABELLE, (1935.)—Age, 6 years. Jersey. Calved September 19, 1877. Greatest yield when fresh, 12 quarts. Yield May 20, 1878, 8 quarts. Made 9 pounds butter in seven days. Tried October, 1877. Milks to one month of calving.	No. 4.—ISABELLE.—Jersey thorough bred. Flanders, 3d. Quantity, 16 quarts. Quality, first rate. Dry one month.
No. 5.—MARIAN.—Age, 6 years. Guernsey. Calved February 15, 1878. Greatest yield when fresh, 14 quarts. Yield May 20, 14 quarts. Butter never been tested. Milks to within one month of calving.	No. 5.—MARIAN.—Guernsey. Curveline, 3d. Quantity, 16 quarts. Quality, first rate. Dry two months to three months.
No. 6.—URANIA, (2793.)—Age, 5 years. Jersey. Calved January 30, 1878. Greatest yield when fresh, 12 quarts. Yield May 20, 12 quarts. Butter never been tested. Milks to one month of calving.	No. 6.—URANIA.—Jersey thorough-bred. Selvage, 2d. Quantity, 14 quarts. Quality, second rate. Dry one month.

No. 7.—FLORENTIA, (3518.)—Age, 4 years. Jersey. Calved January 1, 1878. Greatest yield, 10 quarts. Yield May 20, 10 quarts. Butter never been tested. Milks to one month of calving.

No. 8.—PAUNACUSSING, (5050.)—Age, 2 years. Jersey. Calved October 30, 1877. Yield, May 21, 1878, 8 quarts.

Duration yet to be ascertained.

No. 9.—LADY DELAWARE, (5051.)—Age, 2 years. Thorough-bred Jersey. Calved January 3, 1878. Yield May 21, 1878, 6 quarts.

Duration yet to be ascertained.

No. 7.—FLORENTIA.—Jersey thorough-bred. Curveline, 2d. Quantity, 12 quarts.

Quality, second rate. Dry one month.

No. 8.—PAUNACUSSING.—Jersey thorough-bred. Selvage, 2d. Quantity, 12 quarts. Quality, medium. Dry two months, probably.

No. 9.—LADY DELAWARE.—Jersey thorough-bred. Flanders, 3d.

Quantity, only milks out of two teats. Quality, medium. Dry three months, probably.

Examination of Moses Eastburn's Cow, Beauty.

ACCOUNT OF MOSES EASTBURN.

COW, BEAUTY.—Age, 9 years. Calved March 20, 1878. Greatest yield of milk per day about 17 or 18 quarts. Yield May 24, 1878, 16 quarts. Duration of yield of milk, ten months. Quality of milk, 9 quarts to make a pound of butter. Butter made in eight and a half months, 302½ pounds.

OPINIONS OF THE COMMISSION.

BEAUTY.—Jersey. Curveline, 2d.

Quantity, 18 quarts.

Quality, first class.

Time, well up to her time.

"This is to certify that I was present at the examination of my cow, Beauty, this first of sixth month, 1878, by the committee to test the Guenon system, and find their report to correspond with the within statement.
MOSES EASTBURN."

SOLEBURY, BUCKS COUNTY.

Examination of Colonel James Young's Herd, at Middletown.

The Commission visited the large farms of Colonel James Young, near Middletown, and examined thirty-seven head of cows and heifers, among which were some of the finest Jersey cows in the State. His whole stock is well-fed and cared for, and are in fine condition. He supplies Middletown with the best of milk. Colonel Young does not keep a record of the performances of his cows, and the commision were therefore obliged to examine the cows, and after making their record, to compare it, item by item of each cow separately, with the knowledge of them had by his very intelligent dairy-woman, who has charge of the cows and the milk, and knows their characters as milk and butter producers well; also has a record of the times of calving of all the cows. The estimates of the commission agreed with hers, on all the hundred and eleven points, except nine points, and where they differed, that difference was in two cases on the quality, and in the other cases on the time. The commission attribute their unanimity on this herd, to the careful selection and breeding of Colonel Young, to his good feeding, and the excellent care that the animals have. These points constantly looked after, maintain the excellence of the herd, and as a consequence the escutcheons correspond, for, as the colonel says, "he never saw a good escutcheon without being on a good animal, and never saw a good animal without a good escutcheon."

MIDDLETOWN, *November 1, 1878.*

"We were present when the commission visited our farms, and examined the stock, and we think they judged rightly of it, in nearly every case—we should say within five per cent. of being entirely correct.

We have examined the account to be printed with the original record, and find it to be correct and corresponding.

JAMES YOUNG,
JAMES S. YOUNG."

Examination of the Herd of William Calder, Esq., Harrisburg.

The commission visited one of the farms of William Calder, near Harrisburg. This gentleman has seven farms, containing nine hundred acres, and keeps a variety of stock. On the farm visited, near the reservoir, the commission examined eight head of grade stock, in very good order, on good September pasturage. The dairyman, a very intelligent man, had no record of the exact quantity and quality of the stock, but, as he milked them himself, a knowledge of their general qualities; and upon hearing the decision of the commission upon each cow, assented to the character given of all of them, except on two points: on one as to yield, and on another as to time. It was pleasing to notice the surprise and delight expressed by him at the exhibition, of entire strangers to the herd, of such accurate knowledge of them as the system showed it could give. And he determined to acquire it forthwith.

The commission saw a very fine black grade cow, with the calf by her side a perfect specimen of the Belted stock, though sired by a thoroughbred Jersey bull—to be accounted for only by the fact that the cow had been served by a Belted bull the third time before this one.

Examination of Several Herds near West Grove, Blanketed and Unblanketed, under the Supervision of a Committee of the Experimental Farm Club.

It had been stated by some that the commission used the ordinary means of judging of the value of cows, in addition to the Guenon tests. This was, of course, entirely denied by the commission; and as it was repeated in the public print, the commission, to settle the matter in the minds of candid men, offered to have any number of cows blanketed, so that only their posteriors could be seen, and then judge of their escutcheons, provided a committee should be present at the examination, view it closely, and give a report. Thus pressed, the challenge was accepted, and there was appointed a committee of five of some of the best farmers and dairymen residing near the Experimental farm. It was also understood that any could attend who wished to, and on the day of the examination three of the committee were present, as well as a number of other farmers. The commission examined the first five in the stable, blanketed, then two unblanketed, then two blanketed, and the remaining four unblanketed. The report of this committee is appended herewith. The cows were examined on a farm of Thomas Gawthrop, near West Grove. Afterward a number of cows were examined on several farms in the neighborhood, in the presence of the committee. No longer time was required to form an opinion on the blanketed cows than on the others, and the comparative results can be judged from the accompanying tables.

The commission met them on the day appointed, at the farm of Thomas Gawthrop, and in the presence of the committee (three being present,) and of others, examined seven cows blanketed, and would have examined more, but the committee said it was useless, as they could see, and had full faith that only the escutcheon was considered by the commission. On this farm thirteen head were examined, and the results are herewith given. All then adjourned to the farms of Mark Hughes, Howard Preston, and Everard Conard, and examined other stock in the presence of the committee. The committee's report will be found annexed, thus setting to rest the charge that the commission were examining by any other than the Guenon test.

REPORT OF PENNSYLVANIA COMMISSION. 79

THOMAS GAWTHROP'S ACCOUNT OF COWS EXAMINED AT THOMAS GAWTHROP'S FARM, SEPTEMBER 20.	THE GUENON COMMISSION'S ACCOUNT OF COWS EXAMINED AT THOMAS GAWTHROP'S FARM, SEPTEMBER 20.
No. 1.—VICTORIA.— Grade, Jersey and Durham. Quantity, first. Quality, first. A first-class butter cow, and milks well up to time.	*No. 1.*—VICTORIA.*—Grade, Jersey and Durham. Eight years. Quantity, first. Quality, first. Up to her time.
No. 2.—CECIL.—Grade, Jersey. Quantity, first. Quality, first. First-class for butter. Milks up to time.	*No. 2.*—CECIL.* Quantity, first. Quality, first. Short eight weeks.
No. 3.—NELLIE. Quantity, second. Quality, second. Dry from ten to twelve weeks.	*No. 3.*—NELLIE.*—Demijohn, 1. Quantity, second. Quality, first. Short eight weeks.
No. 4.—LUCY.—Recently purchased. Yields three months from calving thirteen quarts. Quality, first.	*No. 4.*—LUCY.*—Flanders, 2d. Quantity, second. Quality, first. Up to her time.
No. 5.—LILY.—Grade, Jersey. Quantity, second. Quality, first. Milks up to time.	*No. 5.*—LILY.*—Grade, Jersey. Flanders, 2d. Quantity, second. Quality, first. Dry four to six weeks.
No. 6.—BEAUTY.—Jersey. Yields fourteen quarts per day. Quality, first. Milks to within eight weeks of calving.	*No. 6.*—BEAUTY.—Jersey. Five years old. Quantity, third. Quality, second. Dry two months.
No. 7.—DARBY.—Jersey. Quantity, fourteen quarts per day. Quality, first. Almost impossible to turn dry, though never excelling in quantity.	*No. 7.*—DARBY.—Jersey. Flanders, 2d. Quantity, second. Quality, first. Dry four weeks.
No. 8.—STAR.—Grade, three quarter Jersey. Yield with first calf from twelve to fourteen quarts per day, and milks well up to time. Quality, first class.	*No. 8.*—STAR †—Grade, three quarter Jersey. Flanders, 3d. Quantity, second. Quality, first. Dry six weeks. Her Jersey blood helps to overcome some blemishes on her escutcheon.
No. 9.—NORAH. Quantity, first. Quality, first. Dry from eight to ten weeks.	*No. 9.*—NORAH.†—Grade. Quantity, first. Quality, first. Up to her time.
No. 10. SALLIE.‡ Quantity, second. Quality, second. Goes dry eight weeks.	*No. 10.*—SALLIE.—Grade. Twelve years. Flanders, 2d. Quantity, 2d. Quality, second. Up to her time.
No. 11.—DIDO.‡—Grade. Quantity, twenty quarts. Second in quality. Dry from eight to twelve weeks.	*No. 11.*—DIDO.—Grade. Left Flanders. Quantity, first. Quality, second. Dry three months.
No. 12.—MOLLY.‡ Yields about sixteen quarts per day. Second-class quality. Dry from eight to twelve weeks.	*No. 12.*—MOLLY. Imperfect Flanders. Quantity, second. Quality, second. Dry ten weeks.
No. 13. WHITEFACE.‡ Second-class in quantity, fifteen quarts per day. Second quality. Dry about ten weeks.	*No. 13.*—WHITEFACE.—Grade. Curveline, 3d. Quantity, third. Quality, second. Dry ten weeks.

* These five cows were so blanketed, as to show only the escutcheon.
† These cows were also blanketed.
‡ The last four animals were not blanketed, but were driven up and examined by the commission without any apparent reference to any marks, except the escutcheon. T. G.

Mark Hughes' Account of his Cows, September 20.

*No. 1.—*LACTE.

Do not know the quantity of milk and butter per week, but gives very rich milk, and milks up to calving.

No. 2. LAURA.

Quantity, twenty-four quarts milk per day.
Quality, sixteen pounds butter week.
Has never been dry; begins to increase in milk about three weeks before calving, and cannot be turned dry.

*No. 3.—*TOPSY.

Quantity, twenty quarts milk per day.
Quality, makes thirteen pounds butter per week.
Will milk up to calving.

Commissions Account of Mark Hughes' Cows.

*No. 1.—*LACTE.—Jersey thorough-bred. Flanders, second.
Quantity and quality, first rate.

Milks close to calving.

*No. 2.—*LAURA.—Jersey thorough-bred. Demijohn, 1st.
Quantity, first rate.

Quality, first class.
Milks up to calving.

*No. 3.—*TOPSY.—Grade, Jersey. Ten years. Curveline, 1st.
Quantity and quality, first rate.

Milks up to calving.

Howard Preston's Account of His Cows.

*No. 1.—*Grade Durham.
Quantity, second.
Quality, second.
Milks up to her time.

*No. 2.—*Grade Durham.
Quantity, second.
Quality, second.
Dry ten weeks.

*No. 3.—*Common stock.
Quantity, second.
Quality, second.
Dry eight to ten weeks.

*No. 4.—*Grade Durham.
Quantity, third.
Quality, second.
Dry ten weeks.

*No. 5.—*Common stock.
Quantity, second.
Quality, second.
Dry three months.

*No. 6.—*Grade Durham.
Quantity, second.
Quality, second.
Dry eight weeks.

*No. 7.—*Grade Durham.
Quantity, second.
Quality, third.
Dry ten weeks.

*No. 8.—*NELLY.—Grade Jersey.

Quantity, second.
Quality, second.
Dry eight weeks.

*No. 9.—*JESSIE.—Grade Jersey.
Quantity, second.
Quality, second.
Milks up to time.

*No. 10.—*POLLY.—Grade Jersey.

Quantity, first.
Quality, first.
Milks up to her time.

*No. 11.—*LILY.—Common stock.

Quantity, first.
Quality, first.
Dry eight weeks.

Guenon Commission's Account of Howard Preston's Cows, September 20.

*No. 1.—*Flanders, 2d.—Grade Durham.
Quantity, second.
Quality, second.
Up to time.

*No. 2.—*Left Flanders.—Grade Durham.
Quantity, second.
Quality, second.
Dry two months.

*No. 3.—*Gradestock.—Imperfect Flanders.
Quantity, third.
Quality, second.
Dry eight weeks.

*No. 4.—*Grade Durham.—Selvage, 2d.
Quantity, second.
Quality, second.
Dry ten weeks.

*No. 5.—*Native stock.—Flanders, 3d.
Quantity, second.
Quality, third.
Dry three months.

*No. 6.—*Grade Durham.—Flanders, 3d.
Quantity, third.
Quality, third.
Dry six to eight weeks.

*No. 7.—*Grade Durham.—Bicorn, 3d.
Quantity third.
Quality, second.
Time, eight weeks.

*No. 8.—*NELLY.—Grade Jersey.—Flandrine a Gauche.
Quantity, second.
Quality, second.
Time, eight weeks.

*No. 9.—*JESSIE.—Grade Jersey.—Selvage.
Quantity, second.
Quality, first.
Up to time.

*No. 10.—*POLLY.—Grade Jersey.—Flanders, 2d.
Quantity, second.
Quality, second.
Up to time.

*No. 11.—*LILY.—Native stock.—Flanders, 2d.
Quantity, first.
Quality, first.
Dry four to six weeks.

No. 12.—BLUSH.—Grade Jersey.

    Quantity, first.
    Quality, first.
    Dry six weeks.

No. 13.—TOPSY.—Grade Jersey.

    Quantity, second.
    Quality, second.
    Dry three months.

No. 14.—BONNIE.—Common stock.

    Quantity, first.
    Quality, first.
    Milks up to her time.

No. 15.—DAISY.—Common stock.

    Quantity, third.
    Quality, second.
    Dry three months or more.

No. 16.—KATIE.—Common stock.

    Quantity, third.
    Quality, second.
    Dry six weeks.

No. 12.—BLUSH.—Grade Jersey. Curveline, 2d.

    Quantity, second.
    Quantity, second.
    Dry six weeks.

No. 13.—TOPSY.—Imperfect Curveline.—Grade Jersey.

    Quantity, third.
    Quality, second.
    Dry three months.

No. 14.—BONNIE.—Flanders, 1st.—Native stock.

    Quantity, first.
    Quality, second.
    Up to her time.

No. 15.—DAISY.—Flanders, 3d.—Native stock.

    Quantity, third.
    Quality, second.
    Dry six weeks.

No. 16.—KATIE.—Flanders, 2d.—Native stock.

    Quantity, second.
    Quality, second.
    Dry four to six weeks.

JOSEPH PYLE'S STATEMENT OF HIS COWS:

No. 1.—FAWN.

    Quantity, 10 to 15 quarts.
    Quality, very rich milk.
    Dry from four to six weeks.

No. 2.—FANCY.

    Quantity, 16 to 18 quarts.

    Quality, milk very good quality.
    Falls off sooner than most cows, and will go dry eight weeks.

GUENON COMMISSION'S ACCOUNT OF JOSEPH PYLE'S COWS:

No. 1.—RED GRADE COW—8 years.—Flanders, 2.

    Quantity, 14 or 15 quarts.
    Quality, second.
    Dry about ten weeks.

No. 2.—FANCY.—Guernsey. Flanders, 3.

    Quantity, 18 quarts when fresh, and will begin to reduce and stop two months short of her time.
    Quality, first.
    Will go two months dry.

This cow had been previously examined, May 11, at Thos. M. Harvey's farm. Mr. Harvey had since sold her to Mr. Pyle. The following are the two statements at that time:

T. M. HARVEY:

Quality, medium.
Quantity, 16 quarts and fails fast.
Dry three months.

GUENON COMMISSION:

Quality, medium.
Quantity, poor.
Dry two months.

COMMISSION'S ACCOUNT OF MILTON E. CONARD'S COWS:

No. 1.—LILY.—Grade, Guernsey. Bicorn, 1.

    Quantity, about 20 quarts.
    Quality, first.
    Milks up to her time.

No. 2.—FLOYD.—Flanders, 1.

    Quantity, 18 quarts.
    Quality, very good.
    Milks up to her time.

This is a very correct description of my cows, Lily and Floyd.

    M. E. CONARD.

The above examination of our herds of cows, some of which were covered by a large blanket, completely excluding from view every part of the animal except the escutcheon and back part of udder, subjected the commission to the severest test that could be applied; and agreeing, as their estimate of quality and quantity does, with our previously written reports, leads us to think that in the hands of experts it would be a valuable aid in judging the quality of dairy stock.

    THOMAS GAWTHROP,
    EVERARD CONARD,
    HOWARD PRESTON,
    MARK HUGHES,
        Committee.

WEST GROVE, *11 month 7, 1878.*

The undersigned having been present at the examination of Thomas Gawthrop's herd of dairy cows, by the Guenon commission, on the 2d day of 9 month, 1878, am free to say that, although most of the cows were blanketed from horns to tail, their estimate, in a great majority of them, very nearly corresponded with the owners account previously prepared.

M. E. CONARD.

WEST GROVE, PA., *11 month 7, 1878.*

Joseph Pyle would have signed had he been present at the examination. Expresses confidence in the system.

T. G.

Examination of J. & J. Darlington's Cows, October 2d.

The commission visited the herds of Messrs. J. & J. Darlington, October 2, at Darlington station, on Westchester road, Delaware county. These gentlemen make the finest butter and get the largest price in the market. Their dairy is admirably arranged. They have farms of four hundred and eighty acres, and have a herd of one hundred and sixty-seven cows. They had selected about a fair sample of the herd in two lots of cows. The first lot, from No. 9 to 33, was on one farm, and those numbered from 1 to 14 on the other farm. These gentlemen kept no test of the quality of any cow's milk, and have no exact record of the quantity given by any cow; but as they are experienced dairymen, and thoroughly practical men, they knew about what each cow was giving in milk, and about its general quality, and sufficient to pronounce the grade of each cow, whether first, second, or third class. Therefore, in their record they do not give the exact record, as the committee would have desired, so as to compare with their own estimates, but they give the general qualities of the cow, and the two records must be compared from that stand point. Another matter must be taken into consideration. The Messrs. Darlington are liberal feeders, which accounts partly for their rich, tasty butter, and tends to make their cows do full work. A standard of quarts for first, second, and third class, upon which to estimate the qualities of the cows, was agreed upon between the commission and Messrs. Darlington.

J. & J. DARLINGTON'S ACCOUNT.

No 9.—
 Quantity, first.
 Time, six to eight weeks.
 First-class cow.
No 61.—
 Quantity, second.
 Time, six to eight weeks.
 Second class cow.

No. 4.—
 Quantity, third.
 Time, four to six weeks.
 Third class cow.
No. 1.—
 Quantity, first.
 Time, four to six weeks.
 First-class cow.
No. 41.—
 Quantity, first.
 Time, four to six weeks.
 First-class cow.
No. 22.—
 Quantity, first.
 Time, four to six weeks.
 First-class cow.

GUENON COMMISSION'S ACCOUNT.

No. 9.—Grade Durham.—Bicorn, second.
 Quantity, second.
 Quality, second.
 Time, four to six weeks.
No. 61.—Grade Durham.—Imperfect Flanders, third.
 Quantity, second class.
 Quality, second class.
 Time, two months.
No. 4.—Grade Durham.—Flanders, third.
 Quantity, third.
 Quality, second.
 Dry one month.
No. 1.—Grade Durham.—Flanders, sec'd.
 Quantity, second.
 Quality, second.
 Up to her time.
No. 41.—Grade Durham.—Flanders.
 Quantity, first.
 Quality, second.
 Time, six weeks to two months.
No. 22.—Grade Durham.—Flanders, 2d.
 Quantity, first.
 Quality, first.
 Up to time, say four to six weeks.

No. 6.—
  Quantity, third.
  Time, six to eight weeks.
  Third class cow.

No. 7.—
  Quantity, second.
  Time, eight to ten weeks.
  Second class cow.

No. 67.—
  Quantity, first.
  Time, four to six weeks.
  First-class cow.

No. 19.—
  Quantity, third.
  Time, two to three weeks.
  Third class cow.

No. 32.—
  Quantity, third.
  Time, two to three weeks.
  Third class cow.

No. 1.—
  Agrees with the commission.
  Second class cow.
  Dry about two months.

No. 2.—
  Agrees with commission.
  First-class cow.
  Dry four to six weeks.

No. 3.—
  Second class cow.
  Dry about six weeks.

No. 4.—
  Agrees with committee.
  Second class cow.
  Large milker, but fails too soon.
  Dry from six to eight weeks.

No. 5.—
  Large milker.
  First-class.
  Dry six to eight weeks.

No. 6.—
  Agrees with committee.
  Second class.
  Dry three to four weeks.

No. 7.—
  First-class in every respect.
  Best in the herd.
  Dry four to eight weeks.

No. 8.—
  Agrees with committee.
  Good second class.
  Dry four to six weeks.

No. 9.—
  First-class.
  Dry four to six weeks.

No. 10.—
  Agrees with commission's.
  Second class.
  Dry six to eight weeks.

No. 11.—
  First-class.
  Dry about eight weeks.

*No. 6.—*Grade Durham.—Imperfect Flanders.
  Quantity, third.
  Quality, second.
  Dry eight to ten weeks.

*No. 7.—*Grade.—Flanders, with bastard marks.
  Quantity, second.
  Quality, second.
  Up to her time, six weeks.
  Reëxamined, and shows bastard marks.

*No. 67.—*Grade Durham.—Imperfect Flanders.
  Quantity, first.
  Quality, second.
  Dry eight weeks.

*No. 19.—*Grade.—Selvage, third.
  Quantity, third.
  Quality, second.
  Dry eight weeks.

*No. 32.—*Durham.—Flanders, third, partly bastard.
  Quantity, second.
  Quality, second.
  Dry eight weeks.

*No. 1.—*Grade.—Flanders, third.
  Quantity, second.
  Quality, second.
  Dry two months.

*No. 2.—*Grade.—Flanders, second.
  Quantity, first.
  Quality, first.
  Dry four to six weeks.

*No. 3.—*Grade.—Imperfect Flanders.
  Quantity, first.
  Quality, first.
  Dry six weeks.

*No. 4.—*Grade.—Flanders, second.
  Quantity, second.
  Quality, second.
  Dry six to eight weeks.

*No. 5.—*Grade, Durham.
  Quantity, second.
  Quality, second.
  Dry eight to ten weeks.

*No. 6.—*Grade.—Horizontal, first.
  Quantity, second.
  Quality, second.
  Dry four to six weeks.

*No. 7.—*Grade.—Curveline, second.
  Quantity, second.
  Quality, third.
  Dry four to six weeks.

*No. 8.—*Grade, Durham.—Horizontal, first.
  Quantity, second.
  Quality, second.
  Dry four to six weeks.

*No. 9.—*Grade.—Flanders, a Gauche.
  Quantity, second.
  Quality, second.
  Dry six weeks.

*No. 10.—*Grade.—Flanders, second.
  Quantity, second.
  Quality, second.
  Dry six weeks.

*No. 11.—*Grade.—Double selvage.—Some bastard marks.
  Quantity, second.
  Quality, second.
  Dry ten to twelve weeks.

No. 12.—
  Agrees with commission's.
  First-class.
  Dry six to eight weeks.

No. 13.—
  Agrees with commission.
  Third class.
  Dry four to six weeks.

No. 14.—
  Agrees with commission.
  First-class.
  Dry two to three weeks.

No. 12.—Grade.—Imperfect Flanders.
  Quantity, first.
  Quality, second.
  Dry six weeks.

No. 13.—Grade, Durham.—Flanders, third.
  Quantity, third.
  Quality, third.
  Dry six to eight weeks.

No. 14.—Grade.—Flanders, second.
  Quantity, first.
  Quality, second.
  Up to her time.

We were present at the examination of our stock by the Pennsylvania Guenon Commission, on October 2d, and have examined the accounts here rendered, with the original written opinions, and find them to correspond. The accounts were given by both parties without either knowing anything of the accounts of the other.

                (Signed)   J. & J. DARLINGTON.

Having given the results of their work, the commission would now leave the further solution of the problem to the practical dairymen of the State. They, of course, expect that not only their report, but also the correctness of the system, will be criticised; but if this criticism is conducted with a spirit of fairness, and with a view to obtain the truth, they fully believe the result will be favorable.

    By direction of the commission.

                      WILLIS P. HAZARD,
                            Secretary.

ADDENDA.

The appointment of a commission by Governor Hartranft, in 1878, to investigate and verify the theories of M. Guenon in judging and selecting milch cows, has resulted in much good to the agricultural community. The members of that commission, including Mr. George Blight, who acted upon a similar committee in 1853, thoroughly imbued with the accuracy of the system and the desire to extend its usefulness, have continued to explain this mode of selecting cows whenever an opportunity offered. This has been very frequent, and many hundred cows have been examined in public, and the system explained in every section of the country.

It is fortunate that all other modes of judging cows do not militate against M. Guenon's views; they give the judge only a more certain mode, and, if he has had much practice, a nearly infallible one. There are some points which are in full unison with Guenon's views, but do not appear in his work, and may be spoken of as follows:

1st. All bovine animals have on the skin of the back a *quirl in the hair*, which seems to be a sort of dividing line or point between the hair on the front of the animal and that on the hinder portion. This should be found in the center of the ridge of the animal, that is, equi-distant from the head as from the root of the tail, and should be well defined, but of short fine hair. Frequently it is to be seen on the shoulder; when there, coarse hair is generally the accompaniment, and with that, a thick or tough skin, and no great milking qualities, or if much milk is given, it is not for a long time, nor is the milk of rich quality. The heaviest milkers have this mark, usually on the middle of the back, and the richest, with short fine hair. In short, the nearer the middle of the back, and the smaller the quirl and the finer the hair, the most generally will the cow be the better milker and of the richest quality. This mark Mr. Blight and myself have been testing for a long time, and we feel now that we can recommend it as a very good additional point to judge from.

2d. The tail should be long and squarely placed on the animal at the root, and of thin fine quality, with a good curly or corkscrew switch, and the bone of the tail should extend fully down to the knee and as much below it as possible. The horns should be small, waxy, and crumpled inwards and downwards a little. If they are long, they should be thin and sometimes rather flat.

3d. Bulls; the same remarks apply to these. Their hind legs should resemble, as much as possible, those of the cow, with great length between the hoof and the first joint; this indicates their aptitude to beget heifer calves and good milkers.

4th. On raising calves, proper nourishment should be given; if stinted, the inferior parts develop to the injury of the better; the head and horns will be out of proportion to the rest of the body.

The Breeding and Value of well-selected Butter Cows.

We have frequently endeavored to show that one of the most important advantages of Guenon's system is, that it enables every owner of cows to tell the good from the bad cows, and that by weeding out the poor ones, and raising the tone of his herd, he will increase his profits, and if every farmer in the State will do the same, the increased value of all herds, and the increased results in profits, would amount to many millions yearly.

Pertinent to this subject, Mr. J. H. Walker, of Worcester, Massachusetts, the owner of a very choice herd of Jerseys, embracing members of the Alphea, Victor, and Pansy families, has prepared an article on the BREEDING AND VALUE OF BUTTER COWS, which proves, by tables showing the net results of good and bad cows, the theory that good cows will pay better than poor ones as an investment. We digest his remarks as follows:

In New England, a pound of butter can be made for less money than a pound and a half of beef, taking the animals at birth or beginning with animals two years old.

Taking any good herd of Jersey cows, old and young, from the time the heifers first come in milk, and it will average to make two thirds as many pounds of butter per annum as any person in New England can make in pounds of beef, on any herd of any breed.

The beef is worth six to nine cents, and the butter from twenty to forty cents.

Furthermore, every farmer should know what the difference is in the actual value of the different cows he owns, rating their value upon the money he gets for their product.

An ordinary cow will make about two hundred pounds of butter a year. The tables are intended to show what the difference is in the value of different cows for producing butter, taking as a basis the payment of thirty dollars for a cow that will make two hundred pounds of butter per annum, and for different amounts up to six hundred pounds per annum, assuming that the cow will die at twelve years of age. The interest upon the first cost of the cow, and on her product for each year, is compounded at the rate of six per cent. per annum, up to the day it is assumed the cow will die, taking no account of the value of the stock bred from her.

As long as every business is done upon the basis of interest on investments, we must treat the question of values as applied to cows on that basis. This is the only way to accurately prove the difference in value between one cow and another.

Table A.

If the cow cost thirty dollars, the keeping per annum twenty-five dollars, and the butter sells for twenty-five cents a pound, the *profits* on the cows will be as follows, viz:

Paying $30 00 for a 200 pound cow, he will get in ten years, ... $170 00
" 189 97 " 300 " " " 235 08
" 348 86 " 400 " " " 299 89
" 504 39 " 500 " " " 363 11
" 671 61 " 600 " " " 428 39

Table B.

Including interest on all items, a farmer will make on each cow as follows, (made on a basis of twenty-five cents a pound for butter, and twenty-five dollars a year for keeping,) viz:

Paying $30 00 for a 200 pound cow, he will get in ten years, ... $195 73
" 125 00 " 300 " " " 313 06
" 250 00 " 400 " " " 374 15
" 350 00 " 500 " " " 474 52
" 450 00 " 600 " " " 595 91

Table C.

Reckoning the annual cost of keeping at thirty-five dollars, and butter at thirty cents a pound, *reckoning interest* on her cost, and on all receipts from her, a farmer will make on each cow as follows, viz:

Paying $30 00 for a 200 pound cow, he will get in ten years, ... $182 87
"   125 00   "   300   "   "   " 354 78
"   250 00   "   400   "   "   " 483 49
"   350 00   "   500   "   "   " 654 17
"   450 00   "   600   "   "   " 811 59

Table D.

On an annual cost of keeping of fifty dollars, and price of butter at thirty-five cents:

Paying $30 00 for a 200 pound cow, he will get in ten years, ... $95 76
"   125 00   "   300   "   "   " 318 39
"   250 00   "   400   "   "   " 507 46
"   350 00   "   500   "   "   " 744 20
"   450 00   "   600   "   "   " 960 90

Assuming that each cow, costing at two years old the price named in the tables, will die at twelve years old, the actual value of cows to practical farmers, making annually the different amounts of butter named, is shown.

They show what the cow will make in the ten years, and also what a farmer can afford to pay for each cow making the different amounts of butter named. They show the different amounts the farmer, who buys one of each of the cows named, paying the prices named for each of the five, will make on each, provided no interest is reckoned on the price paid for the cow, or on the butter made from her, during ten years.

These figures are certainly startling to any one who has not taken the trouble to examine this subject, much more so to the farmer who never figures carefully, and does exactly as his father did before him, without regard to the altered circumstances that surround him.

The farmer who shakes his head wisely at his more enterprising neighbor, and insists that cows making as much butter as is mentioned in these five tables do not live and never did, should know that the thoroughbred Jersey cows, Jersey Belle of Scituate, of the Victor family, made 705 pounds of butter in twelve consecutive months; that Eurotas, of the Alphia family, made 778 pounds of butter between November 12, 1879, and October 15, 1880, and dropped a heifer calf on November 4, 1880; that Pansy, sired by Living Storm, dam Dolly 2d, sired by Emperor 2d, made in her four year old form 574 pounds of butter in one year; that imported Flora made 511 pounds of butter in fifty weeks; that Countess made 16 pounds of butter on grass only, when fourteen years old. These well-established facts no intelligent, fair-minded man now disputes, and it is confidently believed that many more Jerseys will make as much butter as have any of those mentioned.

The question which at once suggests itself to farmers who are not satisfied with their present animals, is that of capital. The answer is, " admitting the above figures to be correct, I have no capital to pay the high prices demanded for the best Jersey cows, and I must therefore forego that improvement of my herd, which I know I ought to make." Let us see if this is so.

By any process of reasoning, the " bull is half the herd." Each cow contributes to one calf each year half its qualities. The bull contributes to every calf produced in the herd half its qualities. Some horse-breeders will talk only of the excellences of the stallion. Some farmers will talk only of the excellences of the cows. Both are mistaken. The sire and the dam, each contribute to their offspring, on the average, exactly the same proportion of their excellences or defects.

Some bulls are so powerfully organized, as to be able to stamp their qualities, good or bad, on nearly every one of their progeny, as are some cows; but these are the rare exceptions. Each contribute the same, as a rule. No scientific investigator of the breeding problem, or careful breeder, would any sooner select the offspring of a 600 pound butter cow, got by a bull from a 200 pound butter family, than he would a heifer got by a full brother to the 600 pound butter cow from a full sister to the 200 pound butter bull.

Using a bull from a 400 pound butter family, on heifers from a 200 pound butter family, is just as likely to produce heifers that will make from two hundred to four hundred pounds of butter annually, averaging a yield of three hundred pounds; as the using of a bull from a 200 pound butter family on cows of a 400 pound butter family, would be to reduce the yield of some of the heifers to two hundred pounds, and the average to three hundred pounds. The increasing the butter yield of the heifers from a herd of cows one half by using a bull on them from a family or breed that make twice as much, or the reverse, can be relied upon as certainly as any expected result in the most uncertain of all business, namely : that of breeding.

If these statements are correct, what had a farmer better pay for a bull from a 400 pound butter family, to use on his herd of ten 200 pound butter cows, rather than use a bull from a 200 pound butter family?

It may be said that the keeping would cost more, because the higher bred product must be kept better. There is some truth in this, but the better keeping would affect favorably the poorer animals as well, and whatever the extra feed would cost, it would carry the value of the average yield as much above the figures we are making, as the extra feed would cost.

The ten 200 pound butter cows, in ten years would pay a profit of $1,957 30. If the ten cows bred from them, by using the 400 pound butter bull, would make half as much again butter at the same cost, the general product would be increased by one half, and leave the sum to be deducted for keeping the same, for if the two year old 200 pound butter heifer could be raised for $30, so could the better bred one. The profit on each of them, deducting $54 18, cost of cow, will be $484 64—on the ten, $4,846 40, and on the 200 pound butter cows, the profits would be $1,957 30. The advantages reaped by the farmer who has the product for ten years of heifers bred by using the better bull, will be $2,889 10 more than on the 200 pound butter cows.

If he paid for his bull $1,500, and the bull and all his cows died at twelve years old, the farmer would be as well off as he would have been to have used the 200 pound butter bull.

But there is no necessity of paying $1,500 for a 400 pound butter bull. One hundred dollars will buy a Jersey bull, six weeks old, from a 400 pound butter family, and he will be old enough to use in twelve months. The $100 paid for him, at six per cent. compound interest, would amount to $191 61, in eleven years. The profit on ten butter cows making three hundred pounds over the ten cows making two hundred pounds in ten years, being $2,800, by deducting the $191 61 for the bull that produced them, (counting nothing for the 200 pound butter bull, for he is good-for-nothing,) the actual advantage reaped by the farmer with intelligence and enterprise enough to secure the better bull, in the ten years after his heifers come in, is over $2,500 on the butter alone. The animals are counted of no value when twelve years old, as the price got for those living beyond that age would average to pay only for the losses caused by accident to animals before reaching that age. These figures take no account of the skim-milk or buttermilk, for they are nearly the same in either case, and will pay the taxes and for the care of the animals; but there is one very important source of profit that is not reckoned, and that is the extra value of the progeny, which is shown by the following table, to be $17,424 48.

There must be no mistake made in procuring a Jersey bull calf.

Although, as a breed, they are twice to three times as valuable for butter as common cows, yet any farmer who buys or uses a Jersey bull, because he is a Jersey bull, will sorely repent his venture.

Buy a bull only from the very best families of Jerseys. They are cheaper than the gift of an average good one.

The idea that it costs more to keep Jersey cows than common cows, or that Jersey cows will not take on flesh, for beef, as readily as other breeds, is true in one view, and very erroneous in another and more correct one.

What a Jersey eats, beyond a limited amount, increases the quantity and richness of her milk, not her flesh, and the amount of flesh she carries is proportionally less for any extra feed, because it does not make flesh, but increases the butter globules in her milk. Again, any other breed can be readily dried off at any time, and being dry, or giving but little milk, and that of poor quality, they readily take on flesh, but a good Jersey is "dried off" with great difficulty, and herein she greatly excels all other breeds. Hundreds of Jerseys, milking twelve to sixteen quarts at their flush, hold out so evenly, that they will give many more quarts of milk, and of double the richness, in a year, than eighteen to twenty-four quart cows, of other families, that are dry several months of the year.

It is the experience of every breeder of Jerseys that, *being dry*, they will take on flesh as fast, with a given quantity and quality of feed, as other breeds, not exclusively beef producers.

They are not good for beef, simply because they are good for butter.

From Jersey cows, a farmer in New England can make a pound of butter worth thirty-five cents, with a less quantity of food than they now use to make a pound and one half of beef worth nine cents.

If farmers think there is some error in these statements, they will, like sensible men whose prosperity depends upon the result, sit down and figure out the results for themselves.

Those who talk loudest against them, will hold on to a cow in their herd that has a little Jersey blood in her; and if they put a price on her, it will be from half as much again, to double that of the finer formed cow standing beside her, guiltless of having any Jersey blood in her veins.

If there is an animal to be had any better than the bull any one is now using, it ought to be secured at once. So with cows, but by all means change at once for a better, any bull, however good.

It is not claimed for any of the tables herewith presented, that they show absolutely the value of any cow to any farmer, but only that they are relatively correct. Every man who consults them, must make his own adjustments as to cost and receipts on any cow he owns. It is clear, that adding a very little to the cost of keeping, and deducting a very little from the price of butter, will show that any 200 pound butter cow brings her owner in debt, each year. Again, there are probably hundreds of cows kept for the dairy, that will not make two hundred pounds of butter in one year on the same feed Jersey Belle of Scituate, had when she made seven hundred and five pounds of butter in one year. It may be said that no allowance is made for any accidents to which a cow is liable—to abort, to have a calf die at birth, to injury, &c., and the thought is present that the loss on the poorer animal is not so much, in that case, as on the better; but the better is no more liable to such a case, and the loss is nearly the same proportionally. But it is still true, that the nearer to absolute worthlessness animals are, the less the loss, relatively and absolutely, their owner suffers in their injury. Better remember, however, that "blessed be nothing" is not the ejaculation of the healthful, the enterprising, and the successful, but of desperate disease, incapacity, or idleness.

Table E.

Showing the value of the progeny of a herd of 32 cows, that each make 300 pounds of butter annually, at the expiration of ten years, together with the value of the butter the progeny will have made during the ten years. Also showing the same on a herd of 32 cows, each making 200 pounds of butter annually. No account is taken of the bull calves, for they are worth nothing. No one can afford to use a bull, however good, if one is to be had that is any better.

On January 1st, of the year—	The original herd of 32 will drop—	Coming in milk at 2, will make butter—	200 POUND BUTTER HERD.			300 POUND BUTTER HERD.		
			Value of butter at the end of ten years.	Value of heifers at end of ten years.	Total value of the heifers and their product.	Value of butter at the end of ten years.	Value of heifers at end of ten years.	Total value of the heifers and their product.
1881	16 heifers,	8 years,	$2,174 46	$160	$2,334 46	$6,973 12	$1,218	$8,129 12
1882	16 "	7 "	2,724 54	160	2,884 54	6,073 12	1,824	7,897 12
1883	16 "	6 "	2,247 04	320	2,567 04	5,117 92	2,432	7,549 92
1884	16 "	5 "	1,740 32	480	2,220 32	4,104 32	3,040	7,144 32
1885	16 "	4 "	1,202 72	480	1,682 72	3,020 12	3,040	6,060 12
1886	16 "	3 "	632 32	480	1,112 32	1,888 32	3,040	4,928 32
1887	16 "	2 "	208 64	480	688 64	980 48	3,040	4,020 48
1888	16 "	1 "	....	480	480 00	340 32	3,040	3,380 32
1889	16 "	yearling,	....	288	288 00	....	2,000	2,000 00
1890	16 "	calf,	....	96	96 00	....	960	960 00
Product of the Second Generation.								
1883	8 heifers,	6 years,	$1,122 52	160	$1,283 54	$2,558 96	1,216	$3,774 96
1884	8 "	5 "	870 16	240	1,110 16	2,052 16	1,520	3,572 16
1885	8 "	4 "	601 36	240	841 36	1,514 56	1,520	3,034 56
1886	8 "	3 "	316 16	240	556 16	944 16	1,520	2,464 16
1887	8 "	2 "	104 32	240	344 32	490 24	1,520	2,010 24
1888	8 "	1 "	....	240	240 00	170 16	1,520	1,690 16
1889	8 "	yearling,	....	144	144 00	....	1,000	1,000 00
1890	8 "	calf,	....	48	48 00	....	480	480 00
Product of the Third Generation.								
1885	4 heifers,	4 years,	$300 63	120	$420 63	$737 28	760	$1,517 29
1886	4 "	3 "	158 03	120	278 03	472 68	760	1,232 68
1887	4 "	2 "	52 16	120	172 16	245 12	760	1,005 12
1888	4 "	1 "	....	120	120 00	85 08	760	845 08
1889	4 "	yearling,	....	72	72 00	....	500	500 00
1890	4 "	calf,	....	24	24 00	....	240	240 00
Product of the Fourth Generation								
1887	2 heifers,	2 years,	$26 08	60	$86 08	$122 56	380	$502 56
1888	2 "	1 "	....	60	60 00	42 64	380	422 64
1889	2 "	yearling,	....	36	36 00	....	250	250 00
1890	2 "	calf,	....	12	12 00	....	120	120 00
Product of the Fifth Generation.								
1889	1 heifer,	yearling,	....	18	18 00	....	125	125 00
1890	1 "	calf,	....	6	6 00	....	60	60 00

Total value of progeny from herd of 32 in 10 years, ... $21,228 53 .... $76,934 52
    Value of progeny, $2,405 77 on each 300 pound cow.
    Value of progeny, $663 38 on each 200 pound cow.

www.ingramcontent.com/pod-product-compliance
Lightning Source LLC
Chambersburg PA
CBHW020154170426
43199CB00010B/1030